DESIGN OF COLUMN-REINFORCED FOUNDATIONS

MOUNIR BOUASSIDA

J.ROSS PUBLISHING

ISBN-13: 978-1-60427-072-3

Printed and bound in the U.S.A. Printed on acid-free paper.

10 9 8 7 6 5 4 3 2 1

Library of Congress Cataloging-in-Publication Data

Name: Bouassida, Mounir.
Title: Design of column-reinforced foundations / by Mounir Bouassida.
Description: Plantation, FL : J. Ross Publishing, [2016] | Includes bibliographical references and index.
Identifiers: LCCN 2016028550 | ISBN 9781604270723 (hardcover : alk. paper)
Subjects: LCSH: Reinforced soils. | Soil stabilization. | Columns—Foundations. | Trenches--Mathematical models. | Soil dynamics.
Classification: LCC TA711 .B68 2016 | DDC 624.1/51363—dc23 LC record available at https://lccn.loc.gov/2016028550

Direct all inquiries to J. Ross Publishing, Inc., 300 S. Pine Island Rd., Suite 305, Plantation, FL 33324.

Phone: (954) 727-9333
Fax: (561) 892-0700
Web: www.jrosspub.com

CONTENTS

Foreword. ix
Preface. xi
About the Author xv

Chapter 1 An Introduction to Column-reinforced Foundations 1
Abstract. 1
1.1 Ground Improvement Techniques 1
1.2 Reinforcement by Columns 3
1.3 Modeling of Column-reinforced Foundations 5
1.3.1 Isolated Column and Trench Models 5
1.3.2 Unit Cell Model 6
1.3.3 Group of Columns Model. 7
1.4 Installation of Columns 9
1.4.1 Sand Compaction Piles 9
1.4.2 Stone Columns and Vibro-compaction 11
1.4.3 Deep Mixing Method for Lime-Cement Columns. 15
1.5 Conclusion 18
References. 19

Chapter 2 Predicting the Ultimate Bearing Capacity of Column-reinforced Foundations 21
Abstract. 21
2.1 Introduction 21
2.2 Design Methods for Isolated Column and Trench Models. 22

2.2.1 Prior Contributions 22
2.2.1.1 Use of the States of Stresses 23
2.2.1.2 Approaches Based on Failure Mechanisms and States of Stresses 25
2.2.1.3 Approaches Based on Failure Mechanisms 27
2.2.1.4 Floating Columns. 29
2.2.2 Bearing Capacity of an Isolated Column and Soil Reinforced by a Trench in Limit Analysis Framework 32
2.2.2.1 Identification of Minimum Characteristics of Reinforcing Material 32
2.2.2.2 Quasi-Exact Bearing Capacity of Soil Reinforced by a Trench 33
2.2.2.3 Loading Tests: Validation and Calibration of Models. 39
2.3 Ultimate Bearing Capacity of Column-reinforced Foundation by the Unit Cell Model 40
2.3.1 Previous Contributions—In Brief 40
2.3.2 Equivalent and Homogenized Characteristics of Reinforced Soil 40
2.3.3 Assessment of Predicted Characteristics with Experimental Data 42
2.4 Ultimate Bearing Capacity of Column-reinforced Foundation by the Group of Columns Model 47
2.4.1 Direct Approaches of Limit Analysis 47
2.4.2 Floating Columns. 49
2.4.3 Homogenization Method 52
2.5 Conclusion 54
References. 54

Chapter 3 Settlement Predictions of Foundations on Reinforced Soil by Columns 59
Abstract. 59
3.1 Introduction 59
3.2 Classification of Methods for the Prediction of Settlement of Reinforced Soil by Columns 63
3.3 Empirical and Semi-empirical Methods 65
3.3.1 Greenwood's Chart. 65
3.3.2 Priebe's Method 66
3.4 Settlement Prediction by the Unit Cell Model 68
3.4.1 Methods Adopting Linear Elastic Behavior 68
3.4.2 Methods Adopting Elastoplastic Behavior 69

3.5 Settlement Predictions by the Group of Columns Model 71
3.5.1 Variational Stress Approach in Linear Elasticity 73
3.5.2 Rectangular Raft on Column Reinforced Foundation 74
3.5.3 Some Comments 74
3.5.4 Homogenization Approach 77
3.6 Assessment of Settlement Predictions by Scaled and Full-scale Models 78
3.6.1 Oil Storage Tank 78
3.6.2 Storage Facility of Protected Bullets in Mounded Banks in Tunisia 80
3.7 Conclusions 82
Appendix 3.1 Variational Approach in Linear Elasticity 82
References 83

Chapter 4 Novel Methodology for Design of Column-reinforced Foundations 87
Abstract 87
4.1 Introduction 88
4.2 Problem Statement 90
4.3 Bearing Capacity of CRF 91
4.3.1 Case of Stone Columns 91
4.3.2 Case of Deep Mixing Columns 92
4.4 Settlement of Column-reinforced Foundations 94
4.4.1 Settlement of End-Bearing Columns 94
4.4.2 Settlement of Floating Columns 96
4.4.3 Acceleration of Consolidation 98
4.5 Illustrative Case Histories 99
4.5.1 Oil Tank on End-Bearing Sand Compaction Column Foundation (Tunisia) 99
4.5.2 Embankment on Soft Clay Reinforced by Floating Columns ... 102
4.5.3 Trial Embankment in Saga, Japan 105
4.6 Conclusions 110
Appendix 4.1 110
References 111

Chapter 5 The Behavior of Improved Soft Clay by Stone Column Installation 115
Abstract 115
5.1 Introduction 116
5.2 Stone Column Installation in Soft Clay 118

5.3 Simulation of Stone Column Installation in Soft Soils 122
5.3.1 Numerical Simulations Using the Unit Cell Model. 122
5.3.1.1 Numerical Predictions by the Mohr Coulomb Model. 125
5.3.1.2 Numerical Predictions by the Hardening Soil Model. 127
5.3.2 Numerical Simulations Using the Group of Columns Model 130
5.3.2.1 Processed Numerical Predictions 132
5.3.2.2 Parametric Study for Varied Spacing Between Columns 132
5.3.3 Experimental Simulation of Stone Column Installation in Soft Clay. 138
5.3.3.1 Preparation of Clay Specimen 139
5.3.3.2 Expansion of Cylindrical Cavity 140
5.3.3.3 Triaxial Tests. 142
5.3.3.4 Influence of Cavity Expansion Ratio. 142
5.4 Acceleration of Settlement in Post-final Loading. 152
5.4.1 Equations of Linear Poroelastic Behavior. 152
5.4.2 Statement of the Auxiliary Problem 153
5.4.3 Solution of the Poroelastic Problem 154
5.4.4 Settlement Prediction for a Given Loading History 156
5.5 Conclusions 159
Appendix 5.1 Behavior of Normally Consolidated Soft Clays 161
Appendix 5.2 Details of the Poroelastic Model 163
References. 164

Chapter 6 Behavior of Foundations on Soils Reinforced by Columns. . . 167
Abstract. 167
6.1 Introduction 168
6.2 Study of Reinforced Soil Behavior: Selected Case Histories 169
6.2.1 Foundations on Compressible Clays Reinforced by End-bearing Columns 169
6.2.1.1 Oil Storage Tank at Zarzis (Tunisia) 170
6.2.1.2 Oil Storage Tank at La Goulette (Tunisia) 174
6.2.2 Comparison Between Numerical Predictions in Plane Strain and Axisymmetric Analyses 178
6.2.2.1 Predictions by the HSM (Plane Strain Condition) 179
6.2.2.2 Predictions by the HSM (Axisymmetric Condition) . . . 182

6.2.3 Embankment on Compressible Soil Reinforced by Floating Stone Columns: Ghannouche Case History (Tunisia) 187
6.2.3.1 Prediction of the Settlement of Unreinforced Soil (Plaxis Code). 189
6.2.3.2 Numerical Simulation of the Behavior of a Storage Facility on Reinforced Soil 191
6.2.3.3 Numerical Predictions Using the Group of Trenches Model 194
6.2.4 Comparison Between Predictions by the Plane Strain and Axisymmetric Model of Reinforced Soil. 198
6.3 Conclusions 203
Forthcoming Developments 204
References. 204

FOREWORD

It is my honor and pleasure to write a foreword for Professor Mounir Bouassida's new book on ground improvement, focusing on column-reinforced foundations. Although Professor Bouassida and I are from different parts of the world, there are many common topics and concerns regarding ground improvement in Africa and Asia, as well as unique problems on their own in the two continents. With significant advancement concerning equipment and technologies in ground-improvement techniques in the past two decades, the adoption of deep foundations supporting structures such as tall oil and chemical storage tanks is no longer the definite foundation solution. Ground improvement is often considered as the viable option in terms of construction cost and time considerations while still fulfilling the desired foundation performance specifications. In some cases in Singapore and India from my part of the world as illustrative examples, ground improvement has been adopted successfully as the foundation supporting low- to medium-rise residential and industrial buildings.

In view of the above, there have been many books published on ground improvements along with many conferences held on ground improvement and related topics in recent years. To date, there is no book detailing the design and modeling of reinforced soils by columns and Professor Bouassida's book is a valuable and timely addition to the existing literature on column-reinforced foundations. As far as I know, this is the first ground-improvement book written by a researcher from Africa. Although Professor Bouassida studied in Africa and has done research on ground improvement for the past 25 years in Africa, he had extensive research collaborations in ground improvement with researchers outside Africa that included Australia, France, and the USA—just to name a few. As such, this ground improvement book is unique in a sense that it has an African flavor, but yet with strong international applications.

The contents of the book cover all aspects related to foundations on reinforced soil by columns, including considerations on bearing capacity, settlement, acceleration of consolidation, improvement of soil characteristics due to columns installation, and the study of behavior of such foundation showing the performance of floating columns. In addition, a novel investigation of liquefaction mitigation is also discussed. The book is well organized, with the first chapter introducing column-reinforced foundations in such a way that even readers who are not familiar with the topic can get acquainted with it easily. Topics on bearing capacity and settlement calculations, design methodologies, acceleration of soil consolidation, and case studies are then systematically introduced. Another strength of this book is that it presents the methods of investigations comprising analytical, numerical, and experimental techniques, which are indeed necessary in handling complex engineering problems.

Colin C. F. Leung (Professor)
National University of Singapore

PREFACE

Since 1989, I have investigated soils reinforced by columns at the Ecole Nationale d'Ingénieurs de Tunis (ENIT). Such a theme is of interest because of the quasi-automatic use of deep foundations for versatile infrastructure projects in Tunis, Tunisia. The soil profile in this city is characterized by the presence of a very thick soft clay layer that poses serious challenges for civil engineering foundation designers.

My primary action focused on the calculation of bearing capacity of column-reinforced foundations (CRF) using the direct approach of limit analysis (1990–1996). The main finding during this research period consisted in the results obtained by the unit cell, trench, and group-of-columns models.

This initial investigation was followed by the settlement prediction and the acceleration of consolidation of CRF. Those issues were tackled in the linear elastic framework using the group-of-columns modeling and a poro-elastic behavior considering the unit cell model (2003–2005). In parallel, the study of the improvement of soft soil characteristics was found to be of great importance in view of a better understanding of the observed behavior. Various numerical models led to meaningful findings published from 2003 to 2014.

Meanwhile, the homogenization method was found to be quite interesting for the determination of the bearing capacity of CRF (2001–2006), and the design of floating columns (2009).

This set of results (1995–2005), handling both the bearing capacity and the settlement verifications, served to build up a novel methodology for designing CRF. The main benefits of this methodology were the optimization of the treatment cost, and, more importantly, its applicability for all types of soils and columns installation techniques. The methodology was implemented in *Col-*

umns software that has been commercialized since 2009 by the geotechnical engineering consulting firm, SIMPRO.

As different design codes are used in several countries, the major merit of the present book for practitioners consists in the design of column-reinforced foundations by a unified method applicable to stone columns, deep mixing method, and sand compaction piles.

Therefore, this book represents the fruit of extensive research studies developed over the years by the author in collaboration with several investigators from the academic and professional sides, as well as graduate students involved in their Ph.D. and MSc programs at the Geotechnical Engineering Research Team at ENIT (University of Tunis El Manar).

The editing of this book was handled by distinguished specialists in soil-geomechanics and civil engineering. They truly enhanced the quality of writing and presentation of the book content. Amara Loulizi and Essaieb Hamdi (University of Tunis El Manar), Nigel Balaam (University of Sydney), John P. Carter and André Lyamin (University of Newcastle, Australia), Buddhima Indraratna and Sudip Basack (University of Wollongong, Australia), Colin F. Leung (National University of Singapore), Jun Chin Chai (University of Saga, Japan), Mohammed Yousif Fattah (University of Technology, Baghdad), and Anand Puppala (University of Austin, Texas), please accept my sincere thanks for supporting this publication.

In recognition of their devotion and enthusiastic involvement, from which resulted the publication (to date) of 24 journal articles, three paper discussions, software, 64 papers in peer-reviewed international conferences, 15 invited and keynote papers, and a book chapter: I am, first of all, very grateful to Professors Patrick de Buhan and Luc Dormieux (Ecole des Ponts Paris-Tech, France) who in a friendly manner, intensively collaborated during the initial investigations on bearing capacity, settlement, and acceleration of consolidation predictions throughout several successful Tunisian-French research projects.

Second, on the professional side, the development of several numerical models devoted to the study of the improvement of soft characteristics due to stone columns installation was made possible during collaborative and meaningful work with Mr. Jean-Marc Debats (Vibro-flotation group-Solétanche, France)—a master of technical issues.

Third, the research and devotion of several Ph.D.s and MSc candidates aided this book tremendously. To Tunisian associates Mrs. Samia Boussetta, Mrs. Zahra Fessi-Guetif, Mr. Belgacem Jellali, Mr. Béchir Ben Said, Mrs. Souhir Ellouze, Mr. Lassaad Hazzar, Mr. Wissem Frikha, Mrs. Zeineb Bensalem, Mr. Mnaouar Klai, Mr. Mondher Zouabi, and last but not least Mr. Seifeddine Tabchouche (a young Algerian engineer), you all have my utmost respect and best wishes for your suture success.

Fourth, during initial steps and in preparing my research itinerary, I benefited greatly from the knowledge, the support, and the encouragements from well-respected professors at the Ecole Nationale d'Ingénieurs de Tunis (ENIT): Doctors Ahmed Friaa, Foued Kanoun, Taïeb Hadhri, and Mustapha Besbes: you were, and still remain, among the best examples for the development of higher education and scientific research. You sincerely deserve many honors and high recognition.

I also benefited from several invitations to give courses, seminars, and conferences about the methodology of design implemented in *Columns* software and ground improvement techniques. In this regard, I would like to thank Professors Braja M. Das and Ali Porbaha (Former dean of college and lecturer at CSUS, California, respectively) Andrew J. Whittle (MIT, Massachussets), Buddhima Indraratna (University of Wollongong, Australia), Mourad Karray (University of Sherbrooke, Québec-Canada), Isam Shahrour (Polytech Lille, France), Zoubeir Lafhaj (Ecole Centrale de Lille, France), Mohamed Gharib Sakr and Marawan Shahin (University of Tanta, Egypt), Alain Holeyman (Catholic University Louvain-La-Neuve, Belgium), Nor-Edine Abriak (Mines Douai, France), John P. Carter (University of Newcastle, Australia), David Airey (University of Sydney, Australia), Malek Bouazza (Monash University, Australia), and Jianhua Yin (The Hong Kong Polytechnic University).

Finally, the enormous sacrifices and important support provided by my wife during my career played a key role for all the achievements I have attained. Radhia, you deserve all my respect and love.

Mounir Bouassida (Professor)
University of Tunis El Manar

DEDICATION

This book is dedicated to the memory of my Mother, Emna, and Father, Hedi:

"My Lord! Bestow on them Your Mercy as they did bring me up when I was small."

Translation of Verse 24 of Surah Al-Isra' (The Journey by Night) of The Noble Quran.

ABOUT THE AUTHOR

Mounir Bouassida is a professor of civil engineering at the Ecole Nationale d'Ingénieurs de Tunis (ENIT) of the Université de Tunis–El Manar where he earned his B.S., M.S., Ph.D., and doctorate of sciences diplomas, all in civil engineering. He currently teaches soil mechanics and limit analysis courses at ENIT and the Polytechnic School of Tunisia. He is the director of the Research Laboratory in Geotechnical Engineering and has supervised 13 Ph.D. and 29 master of science graduates. His research focuses on soil improvement techniques and behavior of soft clays.

Dr. Bouassida is the co-author of 45 papers in refereed international journals; 126 papers, including 17 keynote lectures from international conferences; and a book related to testing in soil mechanics. He is a member of the editorial committees of the journal Ground Improvement (ICE), the Geotechnical Geological Engineering Journal (Springer), Infrastructure Innovative Solutions (Springer), and International Journal of Geomechanics (ASCE). He is also an active reviewer in several international journals, including the Journal of Geotechnical and Geoenvironmental Engineering (ASCE) and the Canadian Geotechnical Journal.

As a 2006 Fulbright scholar, Bouassida created a novel methodology for the design of foundations on reinforced soil by columns. He was awarded the 2006 S. Prakash Prize for Excellence in the practice of geotechnical engineering. In 2008, Bouassida launched a Tunisian consulting office in geotechnical engineering, SIMPRO. He is a co-developer of the software *Columns 1.01* used for designing column-reinforced foundations. Professor Bouassida held the office of the vice president of ISSMGE for Africa (2005–2009). He benefited from several grants as a visiting professor in the USA (CSUS and MIT), France (ENPC, Paris, Ecole Centrale de Lille, and Ecole des Mines Douai), Belgium (Catholic University of Louvain–La Neuve), and Australia (University of Newcastle).

ABBREVIATIONS AND NOTATIONS

Abbreviations

2D (3D)	:	Two (three) dimensional
API	:	American Petroleum Institute
ASCE	:	American Society of Civil Engineers
BCF	:	Bearing capacity factor
CPT	:	Cone penetration test
CRF	:	Column-reinforced foundation
DMM	:	Deep mixing method
FE :		Finite element
HSM	:	Hardening soil model
IAR	:	Improvement area ratio
ICE	:	Institution of Civil Engineers
LA	:	Limit analysis
LPG	:	Liquefied Petrolium Gas
MC	:	Mohr-Coulomb
RS	:	Reinforced soil
SC	:	Stone column
SCP	:	Sand compaction piles
SPT	:	Standard penetration test
UCM	:	Unit cell model
YDT	:	Yield design theory

Latin characters

A: area of loaded foundation

A_c : total cross section of columns

a: column's radius

b: radius of unit cell model

b_{cr}: thickness of crown

B: breadth, width of footing

B_1: breadth of trench

$c^{+(-)}$: upper (lower) bound of anisotropic cohesion

c_h : coefficient of horizontal consolidation

c_u : undrained cohesion of the initial soil

$C\ (C')$: (drained) cohesion of the initial soil

C_c : cohesion of column (trench) material

C_{hom} : homogenized cohesion

C^{eq}: equivalent cohesion

$<C> = C_{hom}/C$: normalized homogenized cohesion

dA: elementary area (for integration)

D: depth of embedment

$e_{(0)}$: (initial) void ratio

$\underline{e}_x;\underline{e}_y;\underline{e}_z$: unit vectors of Cartesian coordinates

$\underline{e}_r;\underline{e}_\theta;\underline{e}_z$: unit vectors of cylindrical coordinates

$E(E')$: (drained) Young's modulus of the initial soil

E_c : Young's modulus of column material

E_{dm} : Young's modulus of dummy material

E_{app} : apparent Young's modulus of reinforced soil

E_{hom} : homogenized Young's modulus of reinforced soil

E_M : pressuremeter modulus of deformation

$E_{oed\,(c)}$: oedometer modulus of initial soil (resp. of column material)

F_{rs} : factor of safety against failure of reinforced soil

G: shear modulus

H: stratum depth

H_c : length of columns

H_{emb} : embankment height

K_0 : coefficient of pressure at rest

$k_{v\,(h)}$: vertical (horizontal) permeability

$K_b = B_1 / B$: width ratio

$K_c = C_c / C$: cohesion ratio

$K_{pc} = tg^2(\frac{\pi}{4} + \frac{\varphi_c}{2})$: coefficient of passive pressure of column material

L: length of footing

L_k : characteristic length

M: current spatial point

n: factor of vertical stress concentration

n_i: i^{th} component of outward normal unit vector

N: number of blows from SPT test

N_c: number of columns

O : origin point

p: uniform vertical surcharge

$p_{\lim}$: limit pressure from the pressuremeter test

q: deviatoric stress

q_c: tip resistance from cone penetration test

$Q_{(all)}$: (allowable) vertical load

Q_{ult}^{+} : upper bound of ultimate load

Q_{ult}^{-} : lower bound of ultimate load

q_{u28} : unconfined compressive strength at 28 days

q_{ult} : ultimate bearing capacity

r: radius

R: radius of circular raft foundation

s: axis-to-axis spacing between columns

S: deviatoric macroscopic stress

t: time

$tr(\underline{\underline{\sigma}})$: first invariant of stress field

T: macroscopic shear stress

u: (excess) pore pressure

U_h: horizontal degree of consolidation

$V_{(0)}$: (initial) volume

v : velocity

x, y : plan coordinates

z: depth

Symbol characters

α : parametric angle

β : factor of settlement reduction

δ : total settlement

δ_r : settlement of reinforced soil

δ_{ur} : settlement of unreinforced soil layer(s)

δ_r^+ : upper bound settlement of reinforced soil

$\overline{\delta}$: total allowable settlement

$\overline{\delta}_r$: allowable settlement of reinforced soil

ε_a : axial strain

η : improvement area ratio (IAR)

η_{min} : minimum IAR

η_{max} : maximum IAR

φ (φ'): (drained) friction angle of the initial soil

φ_c : friction angle of column material

φ_{hom} : homogenized friction angle

φ_{eq} : equivalent friction angle

$\lambda_{(c)}, \mu_{(c)}$: Lame's coefficients of initial soil (resp. of column material)

$\gamma_{(c)}$: total unit weight of initial soil (resp. of column material)

Γ : reinforcement factor

$\nu(\nu')$: (drained) Poisson's ratio of initial soil

ν_c : Poisson's ratio of column material

σ_h : horizontal stress

$\sigma_{z(v)}$: vertical stress

$\sigma_{ult,c}$: ultimate vertical stress of column material

$\sigma_{ult,s}$: ultimate vertical stress of initial soil

$\sigma_{ult,rs}$: lower bound of ultimate bearing capacity of reinforced soil

$\sigma_{ult,i}^+$: upper bound of ultimate bearing capacity of medium (i)

σ'_m : effective mean stress

$\underline{\underline{\sigma}}$: stress field

(σ, τ) : normal and shear stresses as axis of Mohr-Coulomb plan

$\sigma_{1(3)}$: major (minor) principal component of stress field

Σ_{ij} : component of macroscopic stress field

$\Sigma_{1(3)}$: major (minor) principal stress of macroscopic field

$(d)\Omega_{(c)}$: (elementary) volume of initial soil (column material)

ξ_i : displacement component along "i" direction

1

AN INTRODUCTION TO COLUMN-REINFORCED FOUNDATIONS

ABSTRACT

Column-reinforced foundations are introduced as a set of in situ ground improvement techniques that consist of installed vertical inclusions to enhance the geotechnical properties of weak and highly compressible soils. Focus is given to the most-practiced worldwide techniques: sand compaction piles (SCP), stone columns, and the deep mixing method (DMM). This chapter introduces the mechanical modeling that will be considered throughout the following chapters, which are aimed at the design and analysis of column-reinforced foundations.

Methods of installation, the types of equipment used, expected performance, and the quality control methods of these techniques are described and reviewed.

1.1 GROUND IMPROVEMENT TECHNIQUES

When facing the construction of structures on problematic soils such as soft clays, the possibility to practice an in situ ground improvement must be well analyzed in order to avoid the expensive conventional deep foundation solution. Soil improvement techniques are not usually time-consuming in terms of installation and they remain cost-effective in comparison to pile foundations.

Nevertheless, when a soil improvement technique is being decided on, it is necessary to check, in situ, both its feasibility and whether the targeted advantages of soil treatment are well-obtained. Accordingly, plot tests are often performed to quantify the improvement of initial soil by comparing its pre- and post-treatment behaviors on the basis of in situ testing results, which also serve for the assessment of adopted assumptions and geotechnical parameters in the design process.

On which basis is a soil improvement technique decided? The unimproved (or initial) soil is primarily identified by its gradation curve which remains a key parameter toward deciding which improvement technique is suitable for it (Dhouib and Blondeau, 2005). Depending on the type of soil (i.e., granular or fine) different improvement techniques can be adopted as sketched in Figure 1.1. Note that for soils having an intermediate grain size distribution, several improvement techniques might be adopted (Bouassida, 2011).

Reinforcement of in situ soil can be made by means of rigid inclusions, like micropiles, and flexible and semi-rigid inclusions, such as column reinforcement techniques, which is the focal theme of this book. Reinforcement

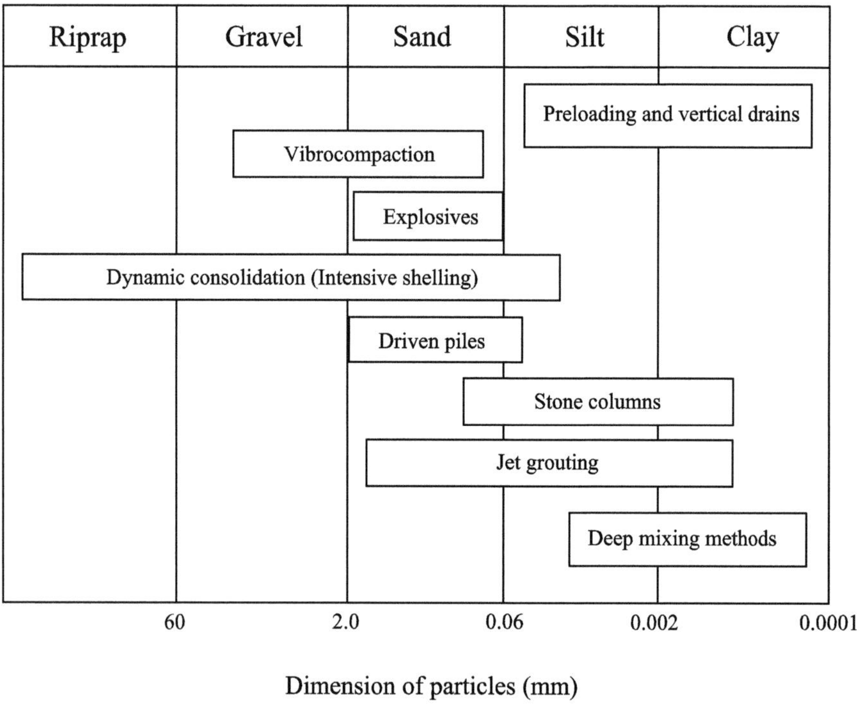

Figure 1.1 In situ ground improvement techniques as a function of original soil grain size distribution (Bouassida, 2011) (with permission from editor A. Bouafia)

of man-made geotechnical structures by transported soil material that is reinforced by added inclusions (e.g., reinforced earth, geogrids, etc.) is not covered.

1.2 REINFORCEMENT BY COLUMNS

The reinforcement by column inclusions of weak and very compressible soils has been practiced worldwide since the 1970s. Among the well-known techniques of installation, we can cite SCP, stone columns, and cement- or lime-treated soil columns.

Column-reinforced foundations usually involve the incorporation of a non-negligible quantity of material in a proportion that varies from 10 to 70 percent of the initial soil volume. The quantity of added material depends on the adopted technique (i.e., stone columns, sand piles, treated soil with cement or lime), as well as the targeted performance.

Reinforcement by columns can be adopted for typically weak soils—hereafter called initial soil, such as highly compressible soft clays—with Young's modulus $E < 3$ MPa and undrained cohesion $c_u < 30$ kPa and loose sands with a friction angle $\varphi < 30°$ (i.e., SPT blow count < 10).

Reinforcement of weak soils by columns aims to increase the bearing capacity, reduce settlement, accelerate consolidation by drained column material, and prevent liquefaction risk—especially in the case of saturated loose sands.

Figure 1.2 presents the load-settlement curves from full-scale field measurements obtained from loading tests of soft Bangkok clay reinforced by sand and gravel piles (Bergado et al., 1984). This figure well illustrates the advantages provided by the reinforcement, namely the increase in bearing capacity: the carrying load is multiplied by three for a settlement of 5 mm; and the very significant reduction in settlement—from 15 mm to approximately 3 mm under a carrying load of one ton.

The cost of a column-reinforced foundation is essentially controlled by the volume fraction of incorporated column material in reference to the improvement area ratio (IAR) that is defined by the total cross section of columns divided by the total area of the loaded foundation.

Table 1.1 summarizes the currently practiced values of IAR as a function of the column techniques and the associated diameters of columns, which are usually assumed to be a circular cross-section for the purpose of design.

Table 1.2 gives the range of enhanced column material rigidity and strength characteristics.

Column-reinforced foundations can be used especially for uniformly loaded structures inducing a surcharge of up to 150 kPa. For example, several types of storage tanks and embankments are built, preferably on reinforced soil, by

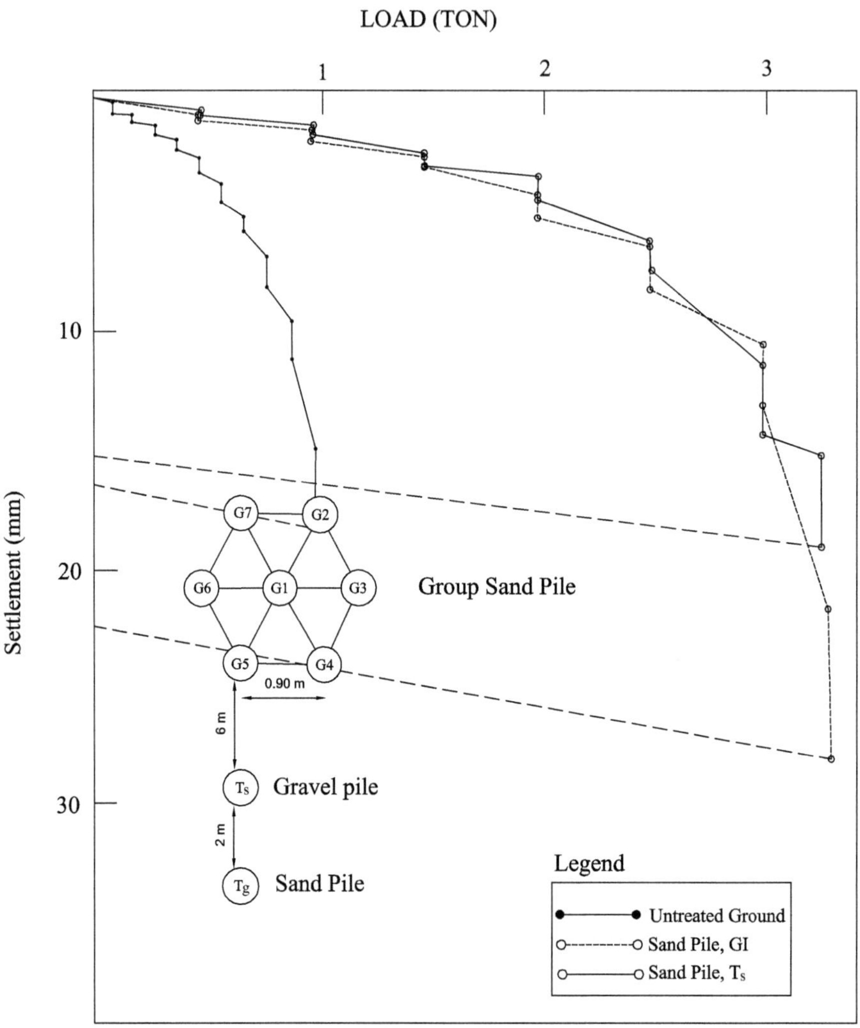

Figure 1.2 Load-settlement relationship of granular piles from full-scale load tests (Bergado et al., 1984) (reprinted with permission from author D. Bergado)

Table 1.1 Practiced IAR and diameters of columns

Columns installation method	Improvement Area Ratio (%)	Columns diameter (m)
Sand compaction piles	$5 < \eta < 10$	0.4 – 0.6
Stone columns & vibro-compaction	$10 < \eta < 33$	0.8 – 1.2
Lime-cement treated soil	$15 < \eta < 70$	0.3 – 0.7

Table 1.2 Mechanical characteristics of column material

Column material	Friction angle	Cohesion, C_c (kPa)	Young's modulus (kPa)
Sand	$35° < \varphi_c < 38°$	0	5 *E* to 10 *E*
Stone & gravel	$\varphi_c > 38°$	5 – 15	15 *E* to 50 *E*
Lime-cement treated soil	$\varphi_c < 20°$	20 *C* – 200 *C*	50 *E* to 200 *E*

end-bearing (i.e., resting on rigid stratum) columns which induce quasi-uniform settlement within an allowable margin.

1.3 MODELING OF COLUMN-REINFORCED FOUNDATIONS

Contributions that are dedicated for the study of column-reinforced foundations have usually considered three mechanical behavior models which can be classified as: the isolated column and trench models, the unit cell (or composite) model, and the group of columns model. Each of the three models is introduced in the following sections.

1.3.1 Isolated Column and Trench Models

The isolated column model is composed of a single inclusion surrounded by an unlimited volume of initial soil. The column is assumed to be cylindrical and, therefore, the axisymmetric condition is suitable for the analysis (Figure 1.3).

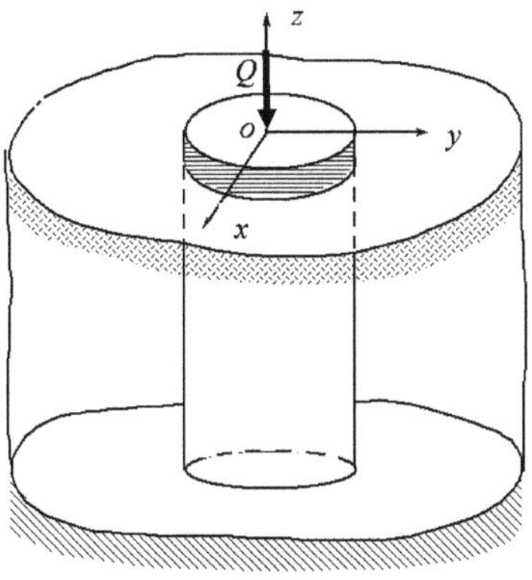

Figure 1.3 Isolated column model (axisymmetric condition) (reprinted from Bouassida, 1996b)

The isolated column is suitable for the assessment of the loading test performed in a trial field on a single column (Bergado and Lam, 1987).

The trench model consists of a wall of reinforcement; it is representative of the plane strain condition (Figure 1.4). The loading is applied on the column or trench cross section which implies that IAR is equal to one (Bouassida, 1996b).

1.3.2 Unit Cell Model

This model is considered from the distribution of a group of columns installed in a regular pattern. Geometrically, it is a reproducible volume which includes a single column with a circular cross section surrounded by a given volume of initial soil. For example, when columns are installed in square or triangular patterns, the periodic volumes of unit cell model correspond to the parallelepiped and hexagonal cylinders respectively, as shown in Figure 1.5.

Then, in order to carry out the axisymmetric condition, an equivalent unit cell with circular cross section is adopted (Figure 1.6). The IAR is written:

$$\eta = \frac{a^2}{b^2} \tag{1.1}$$

The radius of the unit cell model, b, is expressed as a function of the axis-to-axis spacing between columns, s; a denotes the column's radius (Figure 1.5).

The unit cell model is mainly investigated when carrying out laboratory experimental studies in triaxial conditions (Bouassida, 1996a).

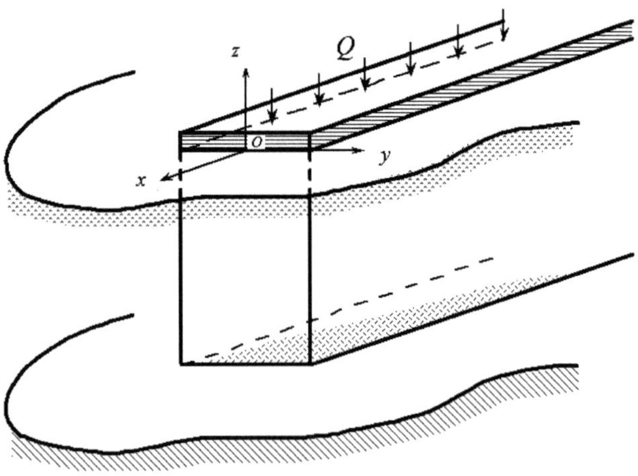

Figure 1.4 Trench model (plane strain condition) (reprinted from Bouassida, 1996b)

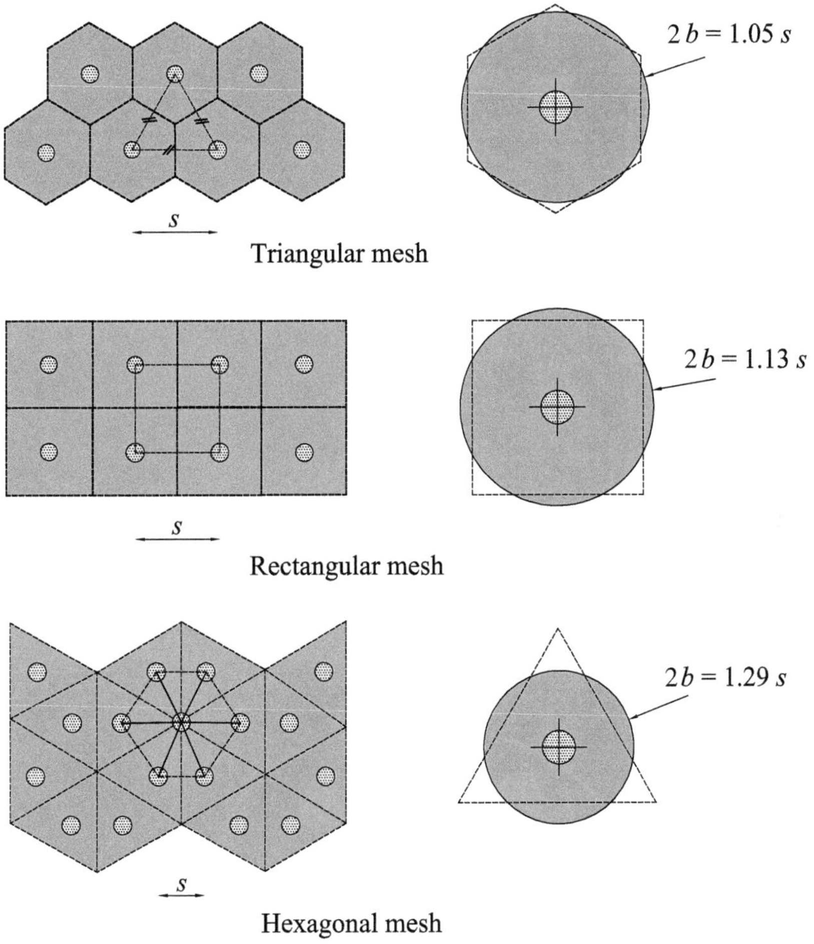

Figure 1.5 Unit cell models as a function of regular columns pattern

1.3.3 Group of Columns Model

The loading structure, whether rigid (rafts or footings) or flexible (embankment), has an arbitrary area (Figure 1.7). All columns having circular cross sections with different diameters are located under the foundation in an arbitrary arrangement. Reinforcement by floating columns represents the general configuration of modeling by assuming total adhesion along the soil-column interfaces. The geometry of columns is defined by their length H_c and total cross section A_c. A rigid stratum is encountered at depth $H > H_c$, (Figure 1.8). The total area of the foundation, denoted by A, is decomposed as:

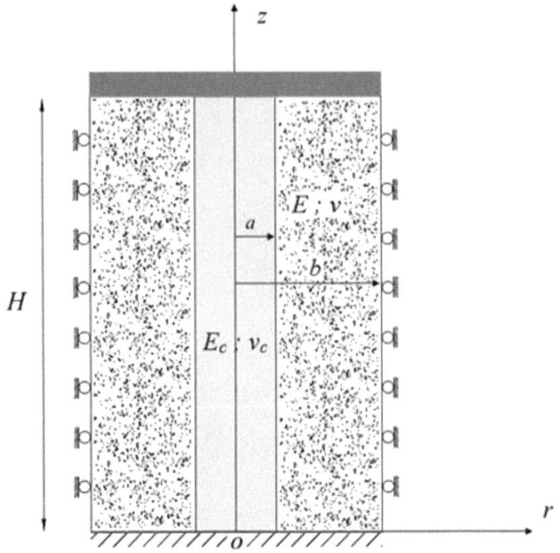

Figure 1.6 Equivalent unit cell model

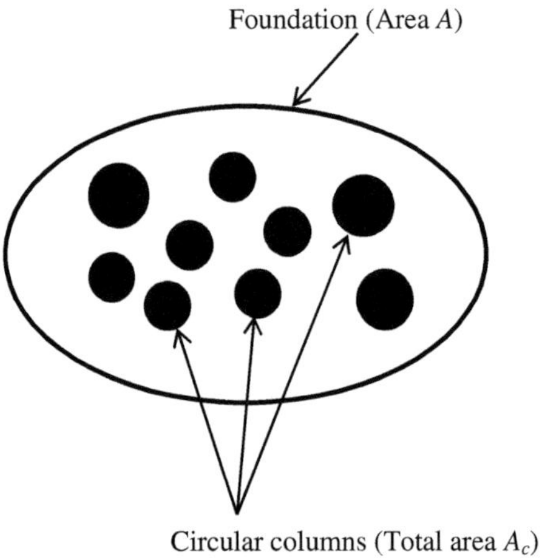

Figure 1.7 Foundation on reinforced soil by a group of columns

$$A = A_c + A_s = \eta A + (1 - \eta) A \tag{1.2}$$

The IAR being defined by:

$$\eta = \frac{A_c}{A} \tag{1.3}$$

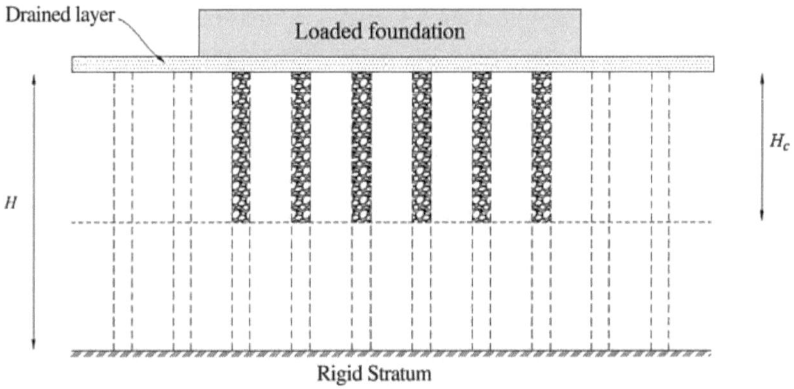

Figure 1.8 Modeling of reinforced soil by a group of floating columns

Geotechnical parameters of reinforced soil constituents E; ν and E_c; ν_c denote the Young's modulus and Poisson's ratio of the initial soil and column material, respectively, which are assumed linearly elastic media. The shear strength characteristics of reinforced soil: C (C_c) and φ (φ_c) denote the cohesion and the friction angle of the native soil and those of the column material, respectively.

Total unit weights of initial soil and column material are denoted by γ and γ_c, respectively.

1.4 INSTALLATION OF COLUMNS

A brief description is given in the following sections about methods of installation, types of equipment used, and illustrations of performances provided by SCP, stone columns, and lime and/or cement treated soils.

1.4.1 Sand Compaction Piles

The SCP method is commonly used to reinforce soft clays and sandy soils especially in coastal areas. Even though it is installed with a low IAR (Table 1.1), the bearing capacity of the composite ground can be increased and its compressibility reduced. The SCP material is granular and has good permeability which makes it behave like a vertical drain by accelerating the consolidation settlement and minimizing the post construction settlements.

Tried in many countries during the last three decades, the SCP method has found very extensive and effective applications, especially in Japan and Korea, for both on-land and marine applications. It consists of well-compacted sand piles that are installed in soft ground by using a vibrating casing pipe (Figure 1.9). Through this process, compacted sand piles are placed by vibration, dynamic impact, or static pressure in the soft or loose ground in a grid pattern over the prescribed area. Installation is done by driving a casing pipe into the soft ground by using a vibratory hammer. The driven casing is then repeatedly filled and compacted with sand. The casing is gradually withdrawn during the filling and compacting, thus creating a dense sand pile.

Among so many contributions, Aboshi et al. (1979) and Kimura (1983) have deeply investigated the mechanical behavior of reinforced ground using the SCP method. These authors discussed the concentration of vertical stress on the surface of composite ground when it is subjected to a loaded rigid raft. Since the settlement in such a case is almost uniform at the surface of the reinforced ground, a vertical stress concentration on the sand columns occurs due to the difference in rigidities of the weak soil and the sand material. The stress concentration factor, defined as the ratio between the vertical stresses induced on the sand piles and soft soil, depends on the IAR. From experimental projects of SCP, the stress concentration factor varied: from 2 to 5 as reported by Vautrain

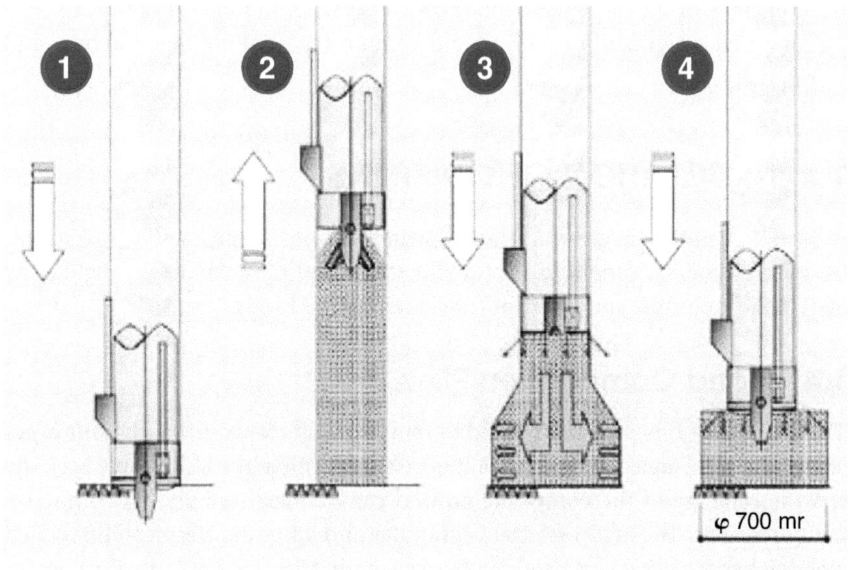

(1): Driving casing pipe into soft ground

(2): Withdrawing casing pipe and bottom-feeding sand pile

(3): Compacting sand pile to increase pile diameter

(4): Completing first cycle of sand pile installation and repeating steps 2 through 4

Figure 1.9 Installation of sand compaction piles

(1980) and Goughnour (1983), up to 9 as reported by Aboshi et al. (1979) and Bergado et al. (1987).

1.4.2 Stone Columns and Vibro-compaction

Stone columns are basically installed using either vibro-replacement or vibro-displacement techniques. More detailed descriptions of the equipment and the procedures used can be found in Moseley and Priebe (1993), Van Impe et al. (1997), Kirsch and Sondermann (2003), etc.

Figure 1.10 shows the possibilities of installing stone columns and vibro-compaction methods for different gradation curves of the original soil. The latter can belong to categories of soils like organic and alluvial clays or gravel and coarse rubble fills (Greenwood, 1970).

Stone columns are often installed in two ways: either by the dry method, which uses (pressurized) compressed air or by the wet method, which uses compressed water. The stages of installation of a stone column using the vibro-displacement technique by the wet method are depicted in Figure 1.11, at a given point of treatment. A vibro-probe (with an extension of tube elements) is gently descended at a prefixed depth under the combined effects of gravity and water pressure. The initial soil is then saturated by induced vibrations to create a tentative localized liquefied zone.

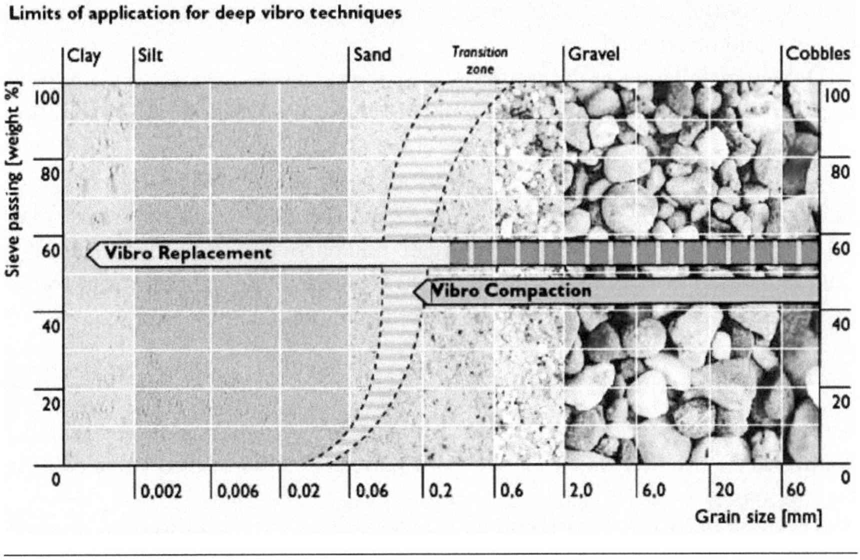

Figure 1.10 Possibilities of installation of stone and vibro-compacted columns as a function of grain size distribution of initial soil (courtesy Keller Gmbh)

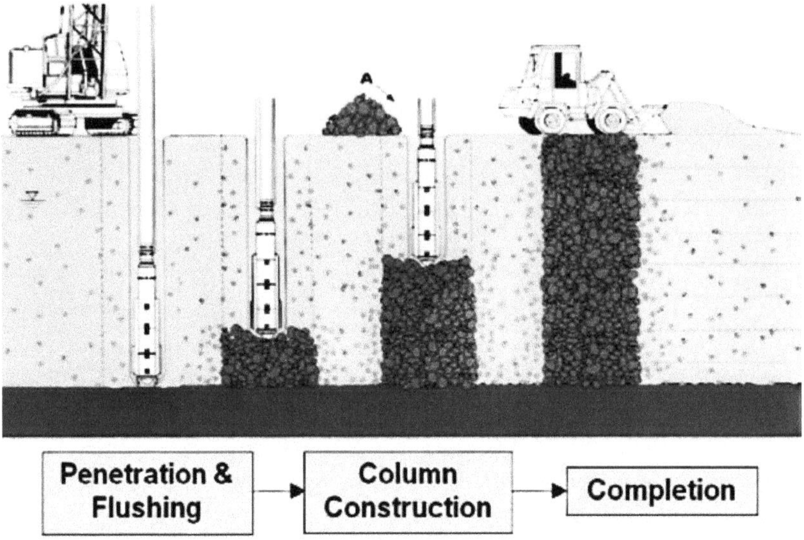

Figure 1.11 Installation of vibro-compacted stone columns, wet method (source from www.keller.com)

Characteristics of the vibro-probe and installation method rely on the type of initial soil and its saturation, as well as on the characteristics of added material and the targeted degree of improvement. The type and grain size of added material should meet standard specifications for homogenous natural stones, either rolled or crushed.

During installation of the columns, the penetration of compacted formations or obstacles might be overcome by a pre-boring, with or without the extraction of initial soil. In such a case, whenever necessary, any excavated volume of initial soil must be filled and compacted by added material. The latter is incorporated until the tip of the vibro-probe is within the created boring, while water is being supplied continuously to evacuate fine particles. The added material is then expanded horizontally, with induced vibrations against the initial soil to form the stone column by successive passes of about 0.5 m in depth.

Control of Installed Columns

Soil reinforcement is conducted with a procedure of quality assurance/quality control to ensure that the targeted performance of each installed stone column is satisfactory.

Monitoring of the installation of a stone column is performed by the use of an automated computerized recorder device. This instrument yields a computer record of the installation process in a continuous graphical mode, plotting depth versus time and compaction effort versus time. The provided information comprises a stone column reference number, date of installation, the required period for installation, and the maximum depth and compaction effort during the penetration and compaction process. These printouts represent the main quality control information of the installation process. Typical stone column monitoring printouts are shown in Figures 1.12 and 1.13 (Debats and Degen, 2001).

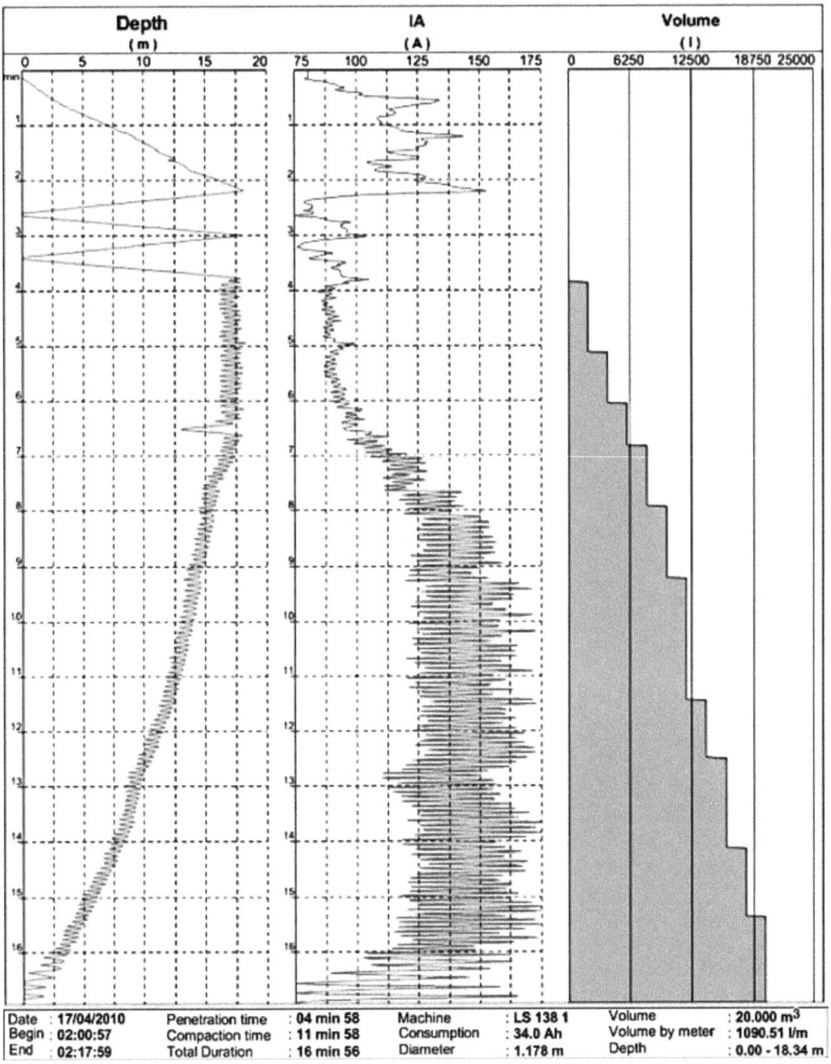

Figure 1.12 Printouts of stone column records for the wet method of installation (courtesy Vibro services; Germany)

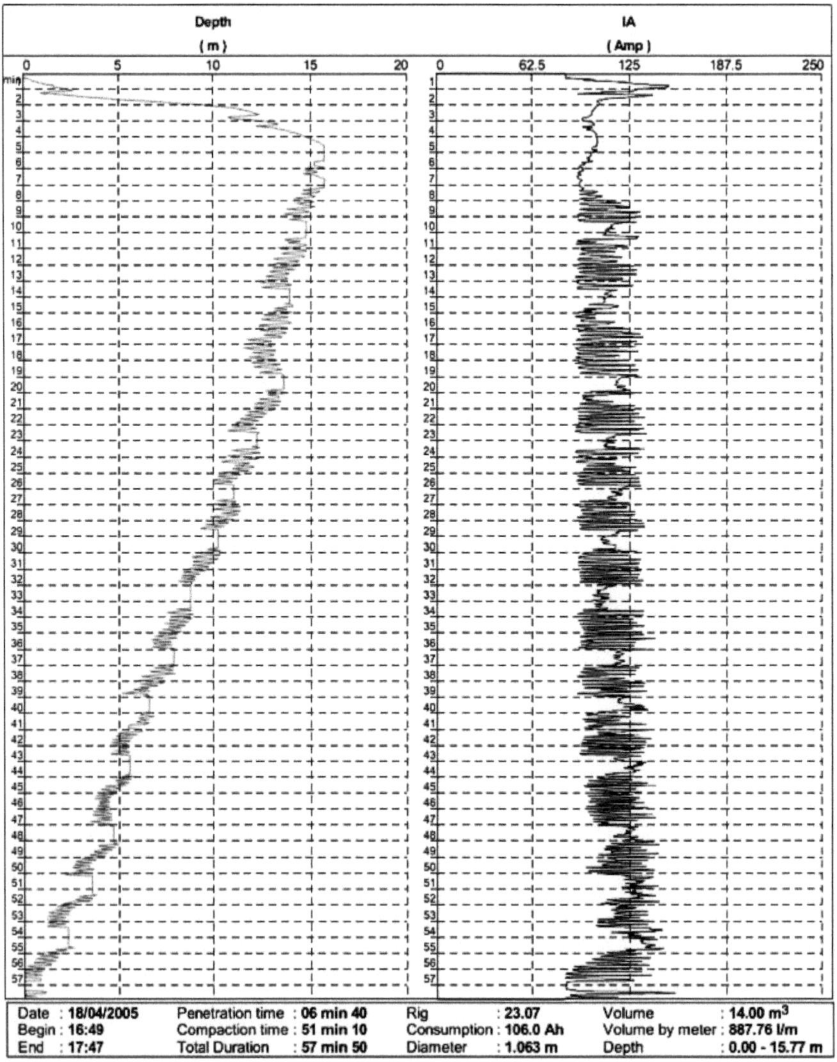

Figure 1.13 Printouts of stone column records for the dry method of installation (courtesy Vibro services; Germany)

Adding to the above-mentioned performances, which result from the installed stone columns, the improvement of initial soil characteristics due to the horizontal expansion of added material is another factor to be considered for the design (Debats et al., 2003).

1.4.3 Deep Mixing Method for Lime-Cement Columns

This technique was primarily tested regarding the reinforcement of soft clays existing around the foundations of harbor facilities and, later on, was extended for embankment and excavation projects.

Field applications of soil-cemented columns had been reported by Bergado et al. (1996) and Porbaha (2000), when they presented, in detail, the practiced process for the treatment of soft soils and performances of the well-known DMM.

Soft clays are usually stabilized by adding a low percentage by weight (8 to 12%) of lime, cement, or lime-cement as a result of pouzzolanic reactions described primarily by Broms and Boman (1977). Since the seventies, the DMM has been extensively practiced in Sweden, Japan, and then worldwide for a wide variety of civil engineering structures.

Steps of installation utilizing the deep mixing technique are shown in Figure 1.14. This technique offers the possibility of installing contiguous columns. It follows that the values of IAR for the reinforcement by the DMM projects are higher than those adopted for the SCP and stone columns applications (Table 1.1).

The choice of binder, such as lime, cement, or lime-cement, depends on several factors: the cost of material, the climate where the treatment is intended, and the targeted gain in shear strength. For example, in the Southern Asiatic environment, Broms (1982) reported that the use of cement is mostly recommended due to the following facts:

- Lime, compared to cement, is not cost-effective;
- Hot-humid climate does not favor the storage of slaked lime; and
- Cement treatment provides higher shear strength.

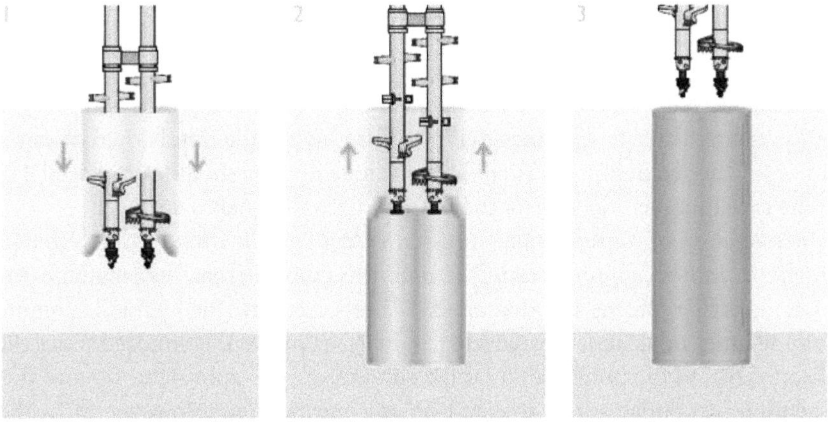

Figure 1.14 Installation using the deep mixing technique (courtesy Keller Gmbh)

Elsewhere, when lime treatment does not provide the due gain in strength, the use of a cement or lime-cement treatment can be decided. Notice that the optimum percentage of lime also depends on the type of initial soil and its grain size curve, more specifically the percentage of fines, i.e., with dimensions less than 2 μm (Hilt and Davidson, 1960) as cited by Bergado et al. (1996).

The gain in shear strength is the main performance provided by cement and/or lime treatments. It is strongly dependent on the curing time during which the pouzzolanic reactions take place between the lime and the existing silicate aluminates in clay. Well-enhanced mechanical characteristics of lime-cement treated soil are usually obtained (Bruce, 2001). As a result, in terms of bearing capacity, full-scale loading tests conducted on isolated lime-treated columns have shown that, depending on the incorporated percentage of lime, the ultimate bearing capacity varies from 50 kN to 500 kN (Broms, 1982).

Several procedures of deep mixing installation have been patented. Among the very recent ones; Trenchmix has been successfully tested by Soletanche-Bachy (2009). Trenchmix is a soil improvement process that enables the installation of a network of trenches as a result of cut in situ soil that is then mixed with an added binder. A spread layer is then overlaid for load transfer to the reinforced soil by the network of trenches. The Trenchmix process is comprised of several steps which start with the placement (preparation) of binder, followed by the mix of ground to be treated by the prepared binder that can be completed by an added fluid. Figure 1.15 shows how the blade is put in a vertical position during the Trenchmix process.

The general procedure of implementation in the dry method, preceded by the provided binder in a pre-excavation, is illustrated in Figure 1.16, which shows a cross section of completely installed Trenchmix. The content of cement used for treatment varies as a function of the type of soil, type of cement, site conditions, and the prescribed unconfined compressive strength of Trenchmix.

Figure 1.17 was derived from records obtained by Soletanche-Bachy (2009) and shows orders of magnitude of cement content and the strength of the treated soil. With respect to performance, the Trenchmix process provides an IAR equaling three to six times that practiced by installed individual columns. However, the maximum spacing between trenches axis should not exceed 4 m, in the case that the thickness of the Trenchmix wall equals 0.4 m.

Acceleration of consolidation—to what extent can the deep mixing technique provide a drained column material? To date, this question remains a matter of discussion. Earlier, Broms and Boman (1977) reported that the hydraulic conductivity of lime- or cement-treated soil can reach as much as a hundred times that of soft clay—to the point that a DMM column can be equivalent to more than a geodrain to enhance the acceleration of consolidation. However, in further

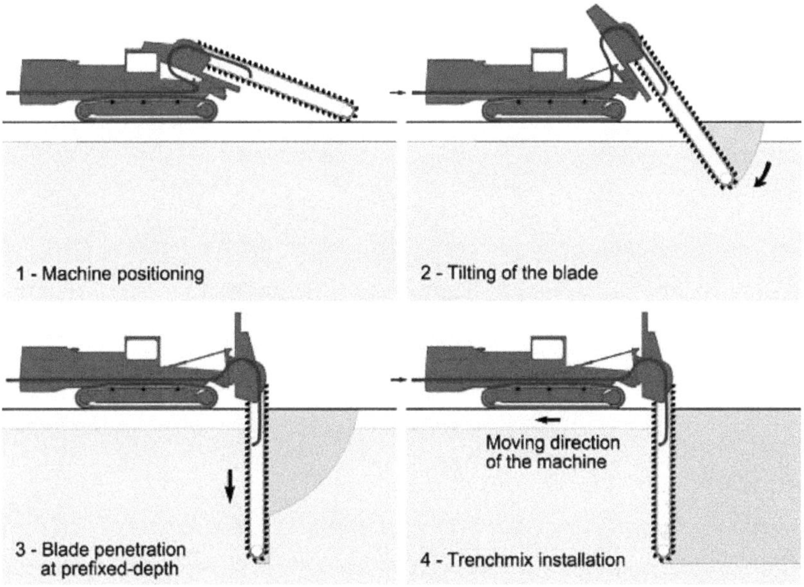

Figure 1.15 Trenchmix process installation (courtesy Soletanche-Bachy)

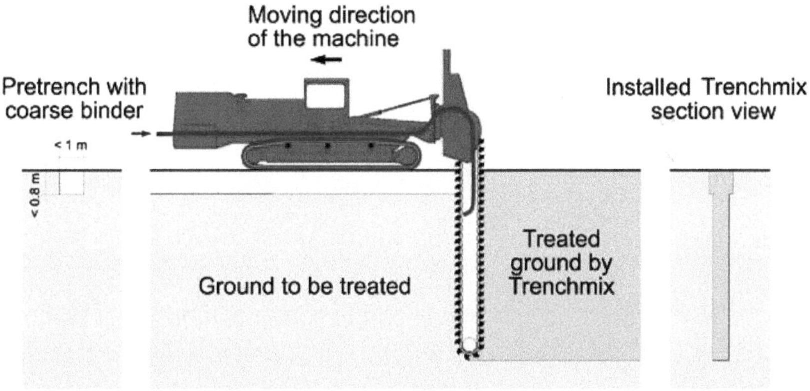

Figure 1.16 Trenchmix implementation procedure preceded by provided binder in a pre-trench (courtesy Soletanche-Bachy)

studies, authors did not fully agree with this observation and stated that the permeability of lime- or cement-treated soft clay can be less or equal to that of unimproved soil. The discussion reported by Chai et al. (2006) has given better insight into the observed acceleration of consolidation of lime-cement-treated

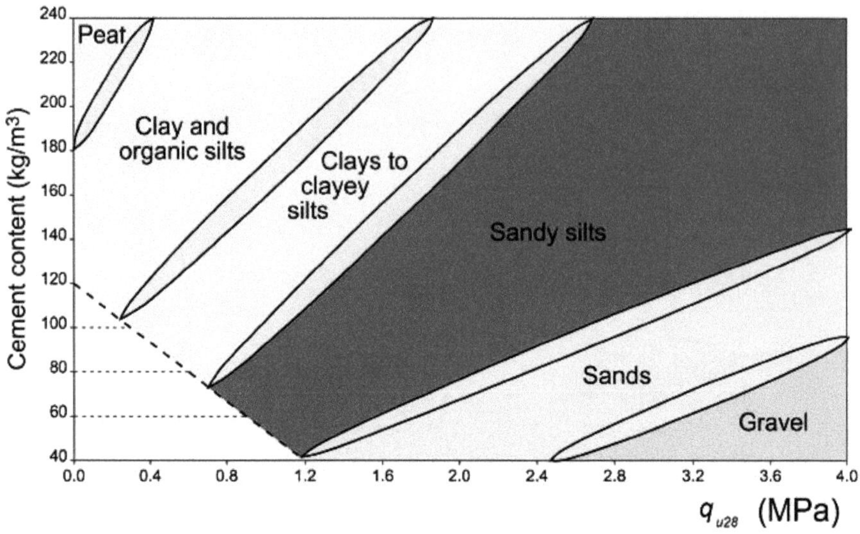

Figure 1.17 Unconfined compressive strength of Trenchmix material versus soil type (courtesy Soletanche-Bachy)

soft soils. Regardless of the magnitude of its hydraulic conductivity, the compression index of treated soil, being proportional to the volumetric modulus (of volume change), is definitely much lower than that of soft clay. Therefore, the significant reduction of compressibility of treated column material will result in the increase of its coefficient of consolidation, which explains the acceleration of consolidation.

1.5 CONCLUSION

An introduction has been given to the techniques of reinforcement by columns which belong to a widespread set of ground improvement methods.

Advantages and methods of installation provided by the column reinforcement techniques have been illustrated. Modeling and specific geotechnical properties of the constituents of reinforced soils by columns were presented in order to carry out all steps of the design, which will be explained in detail in the following chapters.

REFERENCES

Aboshi, H., Ichimoto, E., Enoki, M. and Harada, K., The "Compozer"—a method to improve characteristics of soft clays by inclusion of large diameter sand columns. Proc. Int. Symp. On Soil Reinforcement, Paris, 1, 1979, 211–216.

Bergado, D. T., Anderson, L. R., Miura, N. and Balasubramaniuam, A. S., *Soft Ground Improvement in Lowland and Other Environments.* ASCE Press. New York, 1996.

Bergado, D. T., Huat, S. H. and Kalvade, S., Improvement of soft Bangkok clay using granular piles in subsiding environment. Proc. 5th Int. Geotech. Sem. on case histories in soft clay. Singapore, 1987, 219–226.

Bergado, D. T. and Lam, F. L., Full scale load test on granular piles with different densities and different proportions of gravel and sands on soft Bangkok clay. *Soils and Foundations*, 27 (1), 1987, 86–93.

Bergado, D., Rantucci, G. and Widodo, S., (1984). Full-scale load tests of granular piles and sand drains in soft Bangkok clay. Proc. Int. Symp. "Soil reinforcement," ENPC, Paris, 111–118.

Bouassida, M., Etude expérimentale du renforcement de la vase de Tunis par colonnes de sable—Application pour la validation de la résistance en compression théorique d'une cellule composite confinée. *Revue Française de Géotechnique*, 75 (2), 1996a, 3–12.

Bouassida, M., Determination of bearing capacity of rigid foundations resting on reinforced soil by columns (In French). Doctorate ès-Sciences. Nat. Eng. Sch. of Tunis, 1996b, 246.

Bouassida, M., Amélioration des sols en place. *Introduction à la géotechnique.* Editions de l'Office des Publications Universitaires, Bouafia, A (Coordonnateur.) Algérie, Octobre 2011, Chapitre 6, tome 2, 139–163.

Broms, B. B., Lime columns in theory and practice. Proc. Int. Conf. Soil Mech. Mexico, 1982, 149–165.

Broms, B. B. and Boman, P., Lime columns, a new type of vertical drain. Proc. 9th Int. Conf. on Soil Mechanics and Found. Eng., 2, 1977, 41–75.

Bruce, D. A., An introduction to deep mixing methods as used in geotechnical applications, III: The verification and properties of treated ground. U.S. Department of Transp. Fed. High. Adm. Rep. # FHWA RD, 2001, 99–167.

Chai, J. C., Onitsuka, K. and Hayashi, S., Discussion paper on "Physical modelling of consolidation behavior of a composite foundation consisting of a cement-mixed soil column and untreated soft marine clay" by (Yin, J. H. and Fany, Z. *Géotechnique*, 56 (1), 63–68, 2006, *Géotechnique*, 56 (8), 2006, 579–582.

Debats, J. M. and Degen, W. S., Colonnes ballastées sous la mer—Extension du port de Patras (Grèce). Phase II. Proc. 15th Int. Conf. Soil Mech. and Geotech. Eng., (4), 2001, 2759-2763.

Debats, J. M., Guetif, Z. and Bouassida, M., Soft soil improvement due to vibrocompacted columns installation. Proc. Int. Workshop on Geotechnics of soft soils. *Theory and practice,* Vermeer et al. Eds. September 17–19, Noordwijkerhout, Netherlands, 2003, 551–556.

Dhouib, A. and Blondeau, F., *Colonnes ballastées.* Techniques de mise en oeuvre, domaine d'application, comportement, justification, contrôle, axes de recherche et développement. Presses de l'école nationale des ponts et chaussées, 2005, Paris.

Goughnour, R. R., Settlement of vertically loaded stone columns in soft ground. Proc. 8th Europ. Conf. on S.M.F.E., Helsinki, Finland, 1, 1983, 235–240.

Greenwood, D. A., Mechanical improvement of soils below ground surface. Symp. on Ground Engineering, Institution of Civil Engineers, London, paper II, 1970, 11–22.

Kimura, T., Centrifuge tests on sand compaction piles. Proc. 7th Asian Reg. Conf. S.M.F.E., 1, 1983, 73–122.

Kirsch, K. and Sondermann, W., Ground improvement. Smoltczyk Eds, Geotechnical Engineering Handbook. (2), 1–56, Berlin, 2003, Ernst & Sohn.

Moseley, M. P. and Priebe, H. J., Vibro-techniques Ground Improvement, Edited by M. P. Moseley, Blackie Academic & Professional 1993, 1–19.

Porbaha, A., State of the art in deep mixing technology. Part III: Geomaterials characterization. *Ground Improvement* 4 (3), 2000, 91–110.

Soletanche-Bachy, Trenchmix: Tranchées de sol traité au liant. Cahier des charges particulières. Paris, 2009.

Van-Impe, W. F., De-Cock, F., Van-Der-Cruyssen, J. P. and Maertens, J., Soil improvement experiences in Belgium: part II vibro-compaction and stone columns. *Ground Improvement* J. 1, 1997, 157–168.

Vautrain, J., Comportement et dimensionnement des colonnes ballastées. *Revue Française de Géotechnique,* 11 (2), 1980, 59–73.

2

PREDICTING THE ULTIMATE BEARING CAPACITY OF COLUMN-REINFORCED FOUNDATIONS

ABSTRACT

This chapter focuses on the prediction of the ultimate bearing capacity of column-reinforced foundations (CRF). First, earlier methods of prediction are reviewed, classified, and criticized. Second, the developments made since the 1990s in the framework of limit analysis are summarized in light of the three CRF models presented in Chapter 1: isolated column and trench, unit cell, and group of columns. The limit analysis results obtained by the direct approaches and the homogenization method are interpreted and compared to analytical results predicted by previous methods (Madhav and Vitkar, 1978; Broms, 1982) and recorded test data obtained by Bouassida and Porbaha (2004).

2.1 INTRODUCTION

A review of the methods of design that tried to predict a material's bearing capacity was discussed in detail by Bouassida (1996b). The review indicated the need for a rational framework to properly investigate the bearing capacity for all types of reinforced soil by columns in a determinist manner. In the first stage, the calculation of bearing capacity was addressed using direct static and kinematic approaches of *yield design theory*, better known as limit analysis. Isolated column, the unit cell, and the group of columns modeling have been considered for different types of reinforcement either by end-bearing or by floating columns. The obtained results of bearing capacity have then been assessed with recorded data collected from scaled test models.

The homogenization method has also been used to predict the bearing capacity within the limit analysis framework. New results have been reported to provide solutions for reinforcement cases which were difficult to address using direct approaches of limit analysis, especially in the case of purely cohesive soils reinforced by cohesive-frictional columns. The predicted results by the limit analysis theory have been validated by experimental recorded data, then compared with existing predictions by different existing methods. Reinforcement by floating columns is finally investigated by adopting group of columns modeling.

2.2 DESIGN METHODS FOR ISOLATED COLUMN AND TRENCH MODELS

2.2.1 Prior Contributions

Isolated column and trench models have been considered by many authors who investigated either the analysis of bearing capacity and/or settlement predictions: Bell (1915), Greenwood (1970), Hughes and Withers (1974), Brauns (1978), and Bergado and Lam (1987). Soyez (1985) reviewed and presented most of the suggested methods of design that predict bearing capacity and settlement of column-reinforced soils. At a later time, Bouassida (1996b) classified those methods into three categories, identified their limitations, and suggested possible remediation.

The first category of approaches is based on the study of states of stresses within the initial soil and column material. The second category uses combined states of stresses and failure mechanisms, while the third category considers just failure mechanisms.

Note that trench and isolated column models have not been investigated intensively as compared with the unit cell model (UCM). The earliest contribution of a reinforced soil by trench was presented by Madhav and Vitkar (1987) and Madhav et al. (1979).

2.2.1.1 Use of the States of Stresses

Bell's Approach (1915)

Historically, Bell (1915) made one of the first attempts to use the ultimate vertical stress $\sigma_{ult,c}$ as the bearing capacity of an isolated end-bearing column surrounded by a weightless initial soil subjected to uniform surcharge p (Figure 2.1a). The initial soil is purely cohesive, with a cohesion C, and the column material is purely frictional, with a friction angle φ_c. The shear strength is mobilized within the initial soil and column; the ultimate vertical stress at the column's head is written (Figure 2.1b):

$$\sigma_{ult,c} = (p + 2C)K_{pc} \tag{2.1}$$

$K_{pc} = tg^2\left(\frac{\pi}{4} + \frac{\varphi_c}{2}\right)$ is the coefficient of passive pressure of column material.

The ultimate vertical stress $\sigma_{ult,c}$ in Equation (2.1) is uniformly distributed on the cross section of the column. Therefore, Equation (2.1) applies both for the axisymmetric and plane strain conditions which correspond to the isolated column and trench models, respectively.

Aboshi et al. (1979) had investigated a similar model of reinforced soil as sketched in Figure 2.1, by assuming simultaneous limit equilibrium states that occur within initial soil and column material. They considered the vertical stress $\sigma_{ult,c}$ as a lower bound of the ultimate bearing capacity:

$$\sigma_{ult,c} \leq \sigma_h K_{pc} \tag{2.2a}$$

$$\sigma_h \leq p + 2C \tag{2.2b}$$

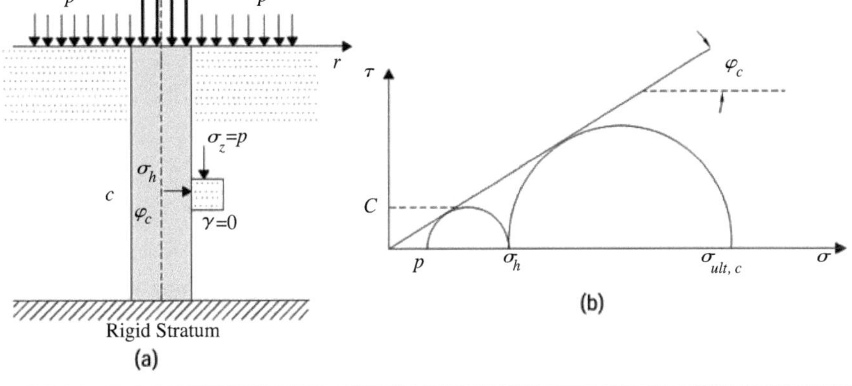

Figure 2.1 Bell's approach (1915). Reprinted from Bouassida (1996b).

Note that the combination of (2.2a) and (2.2b) leads to the result given by Equation (2.1), which confirms the similarity of the Bell (1915) and Aboshi et al. (1979) analyses.

Broms' Approach (1982)

This approach was suggested on the basis of an experimental study on the shear strength of lime treated soft clay. The results obtained from undrained triaxial tests indicated that the shear strength of lime treated soft clay is mainly attributed to the friction that develops between aggregates at low confining stresses. At high confining stresses the observed shear strength develops within lime treated soil aggregates, and provides higher shear strength (Figure 2.2).

Accordingly, Broms (1982) postulated that the ultimate bearing capacity of an isolated column is governed by the shear strength of the stabilized lime rather than the surrounding soft clay. In addition, the bearing capacity depends on the horizontal confining stress as a function of the properties of surrounding soft soil. Broms (2000) suggested that the horizontal confining stress is written:

$$\sigma_h = \gamma z + 5C \tag{2.3}$$

C and γ represent the cohesion and total unit weight of soft clay, respectively.

From the observed undrained and drained behaviors of stabilized soil, Broms (1982) suggested that for low to intermediate confining pressures, the ultimate bearing capacity of an isolated column model is written:

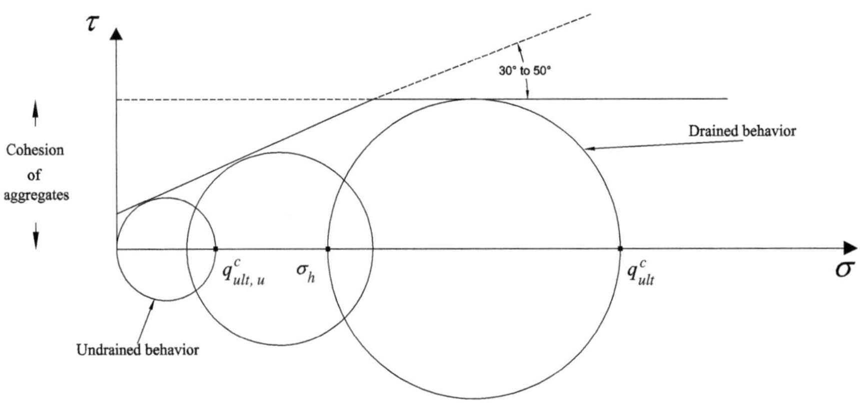

Figure 2.2 Shear strength of lime treated soft clay, (Broms, 1982). Reprinted from Bouassida (1996b).

$$q_{ult,c} = q_{ult,c0} + (\gamma z + 5C)K_{pc} \quad (2.4)$$

$q_{ult,c0}$ is the unconfined compressive strength of the treated column material.

Equation (2.4) provides an empirical estimation of the ultimate bearing capacity of an isolated column since it was suggested from experimental recorded data. Therefore, it cannot be discussed in the framework of limit analysis. Nevertheless, Equation (2.4) needs much more assessment with recorded laboratory or in situ data from studies on lime cement stabilized soft clays. For instance, there is no indication on the limit of applicability of such estimation in terms of depth and magnitude of confinement.

2.2.1.2 Approaches Based on Failure Mechanisms and States of Stresses

Three types of failure mechanisms have been deeply investigated by several authors. Datye (1982) formulated modes of failure which depend on geometrical configurations of isolated columns.

In the case of an end-bearing column with a length that exceeds three to four times its diameter, the failure mechanism by lateral expansion prevails (Figure 2.3). Assuming passive limit equilibrium takes place within the column, Greenwood (1970) primarily expressed the ultimate vertical stress on the column's head as:

$$\sigma_{ult,c} = \sigma_{h,ult} K_{pc} \quad (2.5)$$

Where $\sigma_{h,ult}$ is the ultimate horizontal stress (Figure 2.3) that can be estimated from recorded data, either from laboratory tests or from in situ tests (pressuremeter, cone penetration, or vane test).

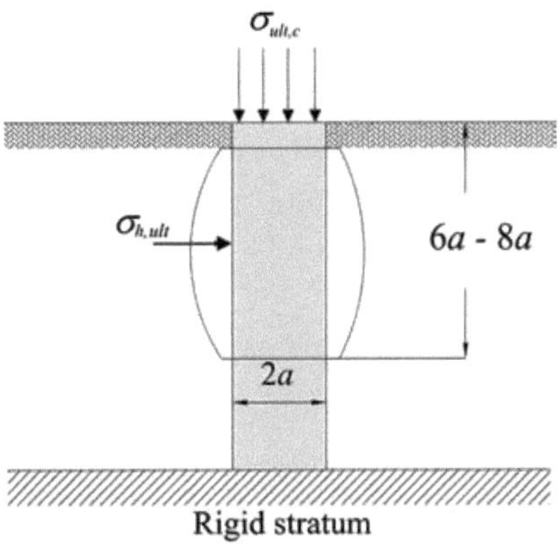

Figure 2.3 Failure mechanism by lateral expansion, Greenwood (1970). Reprinted from Bouassida (1996b).

The failure mechanism by lateral expansion is suitably described by the horizontally expanded cavity model. This failure mechanism was confirmed by Hughes et al. (1975) through experimental analysis carried out on an isolated column. To this end, Hughes et al. (1975) considered perfect elastoplastic behavior and derived the ultimate horizontal stress generated in the initial soil surrounding the column as:

$$\sigma_{h,ult} = \sigma_{h0} + C\left(1 + Ln\frac{E}{2C(1+\nu)}\right) \tag{2.6}$$

Where σ_{h0} is the initial horizontal stress acting at the soil-column interface. E and ν denote, respectively, the Young modulus and Poisson's ratio of initial soil.

Combining Equations (2.5) and (2.6), the ultimate vertical stress on a column's head is written:

$$\sigma_{ult,c} = \left[\sigma_{h0} + C\left(1 + Ln\frac{E}{2C(1+\nu)}\right)\right]K_{pc} \tag{2.7}$$

Later on, a wealth of models were developed in a similar fashion. Some of them use laterally expanded cavity and assume infinite column lengths. Vesic (1972) considered the problem of cavity expansion in cohesive-frictional soils having linear elastic or plastic behavior and suggested the ultimate horizontal stress formula:

$$\sigma_{h,ult} = CF_c' + \sigma_m' F_p' \tag{2.8}$$

σ_m' is the effective mean stress in the initial soil. F_c' and F_p' are dimensionless cylindrical cavity expansion factors which essentially depend on the friction angle of initial soil and its rigidity index defined by:

$$I_r = \frac{E}{2(1+\nu)\left(C + \sigma_m' tg\varphi\right)} \tag{2.9}$$

The ultimate bearing capacity of an isolated column is then deduced from Equation (2.5) where the ultimate horizontal stress is substituted by Equation (2.8).

Consider a purely cohesive initial soil ($\varphi = 0$), then: $F_p' = 0$, the ultimate horizontal stress suggested by Hughes et al. (1975) corresponds to $\sigma_{h0} = 0$ is again deduced from Equation (2.6).

For purely cohesive soils ($\varphi = 0$), that can be assumed incompressible (short term behavior), i.e., $\nu = 0.5$, Equation (2.9) reduces to: $I_r = \frac{E}{3C}$.

Generalized Shear Failure Mechanism, Datye (1982)

In case the rigid stratum is located at a low depth (less than 5 m in practice), the Prandtl's failure mechanism is more likely to happen (Figure 2.4).

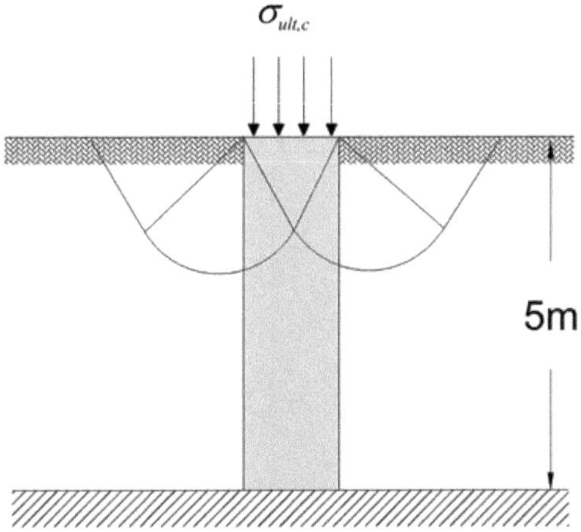

Figure 2.4 Generalized failure mechanism. Reprinted from Bouassida (1996b).

Brauns' Approach (1978)

As investigated by Bell (1915), the ultimate bearing capacity of an isolated column was studied in axisymmetric condition. Passive limit equilibrium is assumed to happen in the column so that the ultimate vertical stress at the column's head is determined from Equation (2.1). Nevertheless, the horizontal stress σ_h at the soil-column interface is derived from a second limit equilibrium equation that reflects the shear strength of initial soil. It is mobilized along the trunk conical surface (IJMI'J'M') of half angle at summit $\left(\frac{\pi}{2}-\alpha\right)$ (Figure 2.5). The ultimate vertical stress at the head of column is a solution of the limit equilibrium equation that is obtained after a semi-analytical minimization over the parametric angle α. The detailed solution related to Brauns' approach has been presented by Bouassida (1996b). Bouassida and Hadhri (1995) mentioned that Braun's approach combines simultaneously the static and kinematic approaches of limit analysis from which the ultimate bearing capacity of the isolated column was found to be overestimated.

2.2.1.3 Approaches Based on Failure Mechanisms

Before the 1980s, approaches essentially based on failure mechanisms were not investigated for the prediction of ultimate bearing capacity of reinforced soils by columns in the framework of limit analysis. The review of existing literature indicates that the contributions by Madhav and Vitkar (1978) and Madhav et al. (1979) seem to be the first to be considered for the trench and isolated column models.

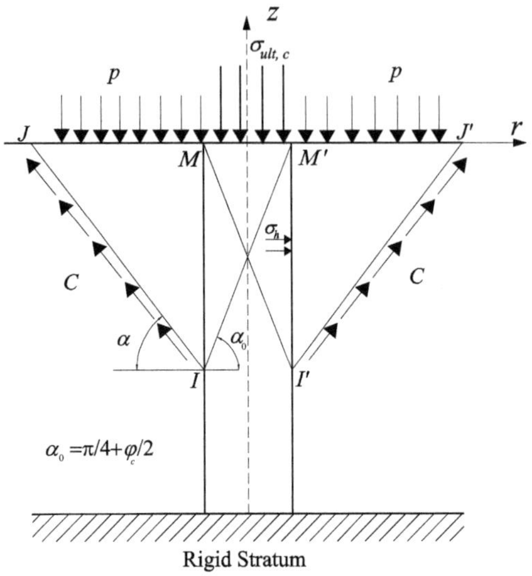

Figure 2.5 Brauns' approach (1978). Reprinted from Bouassida (1996b).

The ultimate bearing capacity of a strip footing, of width B, resting at depth D on reinforced soil by a trench of width $B_1 \leq B$ was studied by implementing a symmetrical Prandtl's mechanism and by taking into account a uniform surcharge load and the gravity of initial soil and column material which are both assumed cohesive-frictional media (Figure 2.6). The failure mechanism is characterized by discontinuities of velocity occurring both along planes (OA), (OA′), (OA″) and (A″A‴) and logarithmic spiral surfaces (AA′) and (A′A″) which bound (OAA′) and (OA′A″) zones in which shear strains are induced. Making use of the kinematic theorem formulated by Drucker and Prager (1952), the ultimate bearing capacity is expressed by:

$$q_{ult} = CN_{cr} + \gamma \frac{B}{2} N_{\gamma r} + \gamma D N_q \tag{2.10}$$

$$N_{cr} = \frac{C_c}{C} N_{cc} + N_c, \qquad N_{\gamma r} = \frac{\gamma_c}{\gamma} N_{\gamma c} + N_\gamma$$

where N_γ, N_c, N_q, $N_{\gamma c}$, and N_{cc} are dimensionless factors which depend on the mechanical characteristics of the initial soil, the column material, the parametric angles α and α_1, and the ratio K_b between the width of the reinforcing trench to that of the strip footing $K_b = \frac{B_1}{B}$ (Figure 2.6). Values of N_{cr}, N_q, and $N_{\gamma r}$ factors are determined after numerical minimization of Equation (2.10) with respect to the angles α and α_1.

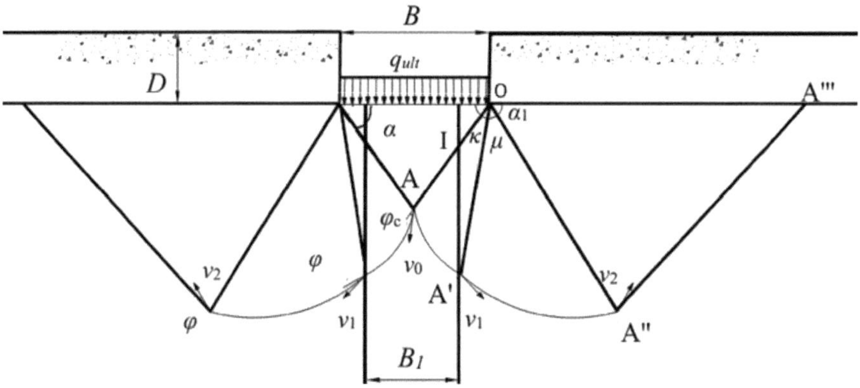

Figure 2.6 Failure mechanism of reinforced soil by a trench (Madhav and Vitkar, 1978). Reprinted from Bouassida (1996b).

Madhav et al. (1979) suggested a modified Hughes' approach that enables the prediction of the ultimate bearing capacity, in axisymmetric condition, of a circular foundation of diameter *B* greater than that of isolated column's $2a$:

$$q_{ult} = \left[(4C + \sigma'_{h0} + K_0 p)\left(\frac{2a}{B}\right)^2\right] K_{pc} + \left[1 - \left(\frac{2a}{B}\right)^2\right] q_{ult,s} \tag{2.11}$$

Where K_0 is the coefficient of lateral pressure at rest of the initial soil, σ'_{h0} is the initial horizontal stress applied over the length of the column, $q_{ult,s}$ is the ultimate bearing capacity of the purely cohesive initial soil: $q_{ult,s} = \frac{2}{3} CN_c$, and p is the uniform vertical surcharge applied at the surface of the initial soil.

Bouassida (1996b) discussed the assumptions made by Madhav et al. (1979) with an emphasis on the surface of discontinuities of velocity occurring especially at the interface between the trench and initial soil media with both obeying Mohr Coulomb's strength criterion.

Table 2.1 summarizes the suggested approaches for the prediction of the bearing capacity of column-reinforced foundations. One can see that the use of limit analysis associated with the homogenization approach makes it possible to solve the problem of bearing capacity when the column-reinforced foundation is modeled as a three-dimensional medium.

2.2.1.4 Floating Columns

Brauns (1978) analyzed the reinforcement of purely cohesive soft soil by an isolated floating column made up of a purely frictional material. Assuming the

Table 2.1 Contributions to the prediction of ultimate bearing capacity of column-reinforced foundations

Modeling	Isolated column/ trench	Unit cell	Group of columns
State of stresses	Bell (1915) Aboshi et al (1979) Broms (1982) Bouassida and Hadhri (1995)	Bouassida et al. (1995); static and kinematic approaches	Bouassida et al. (1995): static approach
Failure mechanism "Limit analysis" (*) Homogenization approach	Madhav and Vitkar (1978) Madhav et al. (1979) Bouassida and Jellali (2002)	Vesic (1972) Hughes and Withers (1974) Aboshi et al. (1979)	Bouassida (1996b): Kinematic approach (*) Jellali et al. (2005) (*) Jellali et al. (2007) (*) Jellali et al. (2011)
Combined failure mechanism and states of stresses	—	Barksdale and Bachus (1983)	Broms (1982) Barksdale (1987)
Experimental investigations	Bergado et al. (1984) Bergado and Lam (1987)	Bouassida (1996a) Ambily and Ghandi (2007)	Bouassida and Porbaha (2004)

column as a rigid pile, from the limit equilibrium of the vertical forces (see Figure 2.7), the vertical stress within the column at depth z is:

$$\sigma_v(z)=\sigma_{ult,c}+\left(\gamma_c-\frac{2C}{a}\right)z \tag{2.12}$$

Where $\sigma_{ult,\,c}$ is the ultimate vertical stress which is assumed to be uniform at the top of the column with radius a. By setting $\sigma_v(z) = 0$ from Equation (2.12), one obtains the minimum length of the column, that is, the depth along which the reinforcement is necessary:

$$H_{c,\min}=\frac{\sigma_{ult,c}}{\dfrac{2C}{a}-\gamma_c} \tag{2.13}$$

Depending on the strength characteristics of the column's material that can be either purely frictional ($C_c = 0$) or purely cohesive ($\varphi_c = 0$), the ultimate vertical stress from Equation (2.12) is respectively: $\sigma_{ult,\,c} = 2CK_{pc}$ or $\sigma_{ult,\,c} = 2C(1 + K_c)$. $K_c = C_c/C$ is the ratio between the cohesion of the column material to that of the initial soil. Substituting these expressions into Equation (2.13) implies that:

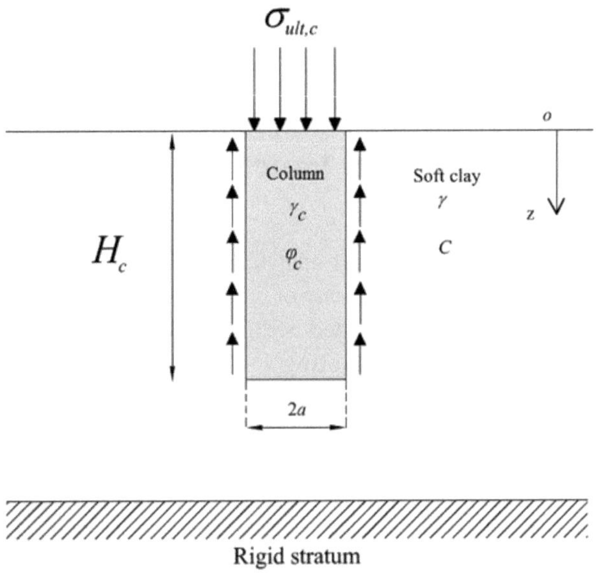

Figure 2.7 Equilibrium of a floating column assumed as a rigid pile (Brauns, 1978). Reprinted from Bouassida (1996b).

$$H_{c,\min} = \frac{2K_{pc}}{\frac{2}{a} - \frac{\gamma_c}{C}} \tag{2.14}$$

$$H_{c,\min} = \frac{2(1+K_c)}{\frac{2}{a} - \frac{\gamma_c}{C}} \tag{2.15}$$

In Section 2.4, predictions by Equations (2.14) and (2.15) of the minimum length of column are compared to those given by methods using the group of columns modeling.

Note that assuming the reinforcing column as a rigid pile is unrealistic particularly for the installation of laterally expanded stone columns or vibro-compacted columns. Therefore, considering the friction resistance at the interface between the column and surrounding soil makes little sense.

In the following, the determination of ultimate bearing capacity of CRF will be considered in the framework of limit analysis. Then, constituents of CRF, i.e., initial soil and column material, are governed by Coulomb's strength criterion that is written:

$$\sigma_1(1-\sin\varphi_{(c)}) - \sigma_3(1+\sin\varphi_{(c)}) - 2C_{(c)}\cos\varphi_{(c)} \leq 0 \tag{2.16}$$

In Equation (2.16) σ_1 and σ_3 denote the major and minor principal stresses which are taken as positive in compression, and negative in tension.

2.2.2 Bearing Capacity of an Isolated Column and Soil Reinforced by a Trench in Limit Analysis Framework

By adopting the framework of limit analysis, detailed in Salençon (1990), the first results of ultimate bearing capacity of reinforced soil by an isolated column and trench models have been reported by Bouassida and Hadhri (1995). Loading parameters comprise a rigid footing subjected to a prescribed downward displacement and gravity of the constituents of reinforced soil. The loaded area by the rigid foundation can be equal or greater to that of the reinforcing trench.

2.2.2.1 Identification of Minimum Characteristics of Reinforcing Material

Bouassida and Hadhri (1995) proposed lower and upper bounds of the ultimate bearing capacity mainly for two cases of reinforcement. They identified the minimum characteristics of reinforcing column (or trench) material from which the increase in bearing capacity (with respect to the bearing capacity of unreinforced soil) occurs. As an example, the reinforcement of a purely cohesive initial soil by a head-loaded trench starts with a purely frictional reinforcing material having an internal angle of friction of 16 to 17 degrees. Elsewhere, the influence of gravity was found to be significant when the friction angle of reinforcing material exceeds 40 degrees. Conversely, the study showed that gravity has an insignificant effect on bearing capacity when the initial soil and trench material are both purely cohesive (Bouassida and Hadhri, 1996). The investigation of the isolated column model has shown that Brauns' approach leads to a significant overestimated bearing capacity of CRF. Nonetheless, the ratio of the loaded area to that of the reinforcing column has a big influence on the lower and upper bounds of the ultimate bearing capacity of CRF.

The trench model was analyzed by Bouassida and Jellali (2002) in the limit analysis framework by assuming that the initial soil and trench material are both cohesive-frictional and have different unit weights. In this context, the width of the rigid foundation can be lower, equal, or greater than the width of the trench. A generalized Prandtl's failure mechanism has been implemented to derive analytically the upper bound of ultimate bearing capacity of CRF (Figure 2.8).

Upper bounds obtained by Bouassida and Jellali (2002) have been compared to those suggested earlier by Madhav and Vitkar (1978) using weightless material with equal widths of rigid foundation and trench. The two methods

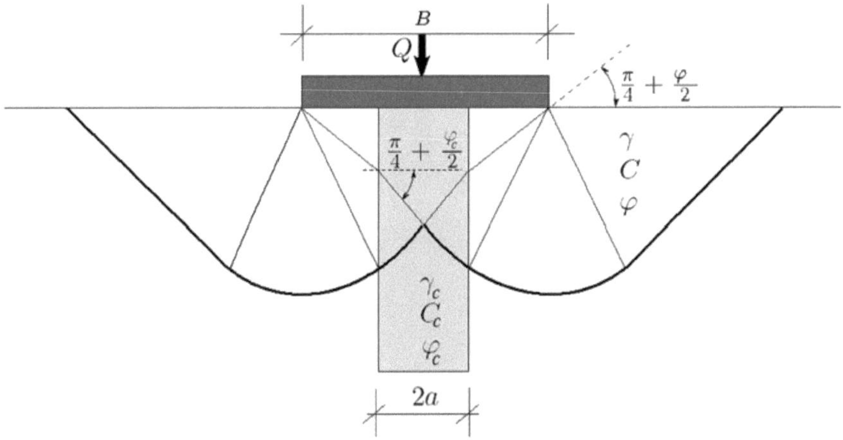

Figure 2.8 Generalized Prandtl's failure mechanism for the trench model (Bouassida and Jellali, 2002)

delivered identical ultimate bearing capacities of reinforced soil by a trench. Notice that the solution primarily suggested by Madhav and Vitkar (1978) is considered to deliver an exact value.

In a case where the gravity of reinforced soil constituents was neglected, Bouassida and Hadhri (1995) performed a five-zone stress field (Figure 2.9), which resulted in the first bearing capacity factor (BCF) lower bound solution of reinforced soil by an end-bearing trench. Considering zero uniform load at the surface of reinforced ground, the lower bound solution is written:

$$Q^{-}/CB=2\left[(1-K_b)\sqrt{K_p}(1+K_p)+K_b(K_{pc}\sqrt{K_p}+K_c\sqrt{K_{pc}})\right] \tag{2.17}$$

In Equation (2.17), the ratio K_b, between B_1 (the breadth of the reinforcing trench) and B (the breadth of the loaded strip footing), represents the improvement area ratio.

2.2.2.2 Quasi-Exact Bearing Capacity of Soil Reinforced by a Trench

Experimental investigations on reinforced soil by a trench were not a matter of high interest as opposed to the numerous contributions developed by means of the UCM. In this way, the validation of the predictions by limit analysis of the ultimate bearing capacity of reinforced soil by a trench is oriented to numerical studies which are of current use, especially when the plane strain condition can be adopted. Therefore, analytical predictions obtained by the trench model will be compared to numerical results which have resulted from the implemented

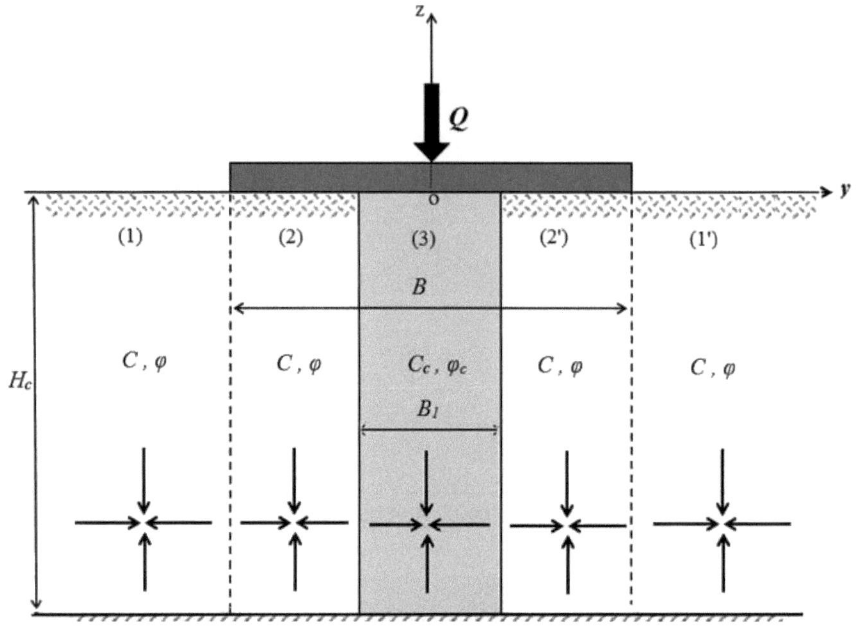

Figure 2.9 Five-zone stress field (Bouassida and Hadhri, 1995). Reprinted from Bouassida (1996b).

Finite Element Limit Analysis (FELA) code developed at the Centre of Geomechanics at the University of Newcastle (Australia).

Two current cases of reinforcement n°1 and n°2 of a purely cohesive soil (C = 20 kPa) have been examined by considering the plane strain numerical model of loaded rigid strip footing resting on reinforced soil by a trench with ratio K_B = 1/3 (as shown in Figures 2.8 and 2.9). Case n°1 corresponds to reinforcement by cohesive-frictional trench material, while case n°2 corresponds to reinforcement by purely cohesive trench material.

A finite element mesh automatically generated by the FELA code has been performed from which the ultimate lower and upper bounds of the BCF have been predicted. Table 2.2 summarizes the parameters of initial soil and trench material considered for the two cases of reinforcement.

Lower and upper bounds of the ultimate BCF are respectively predicted from the five stress field zones (Figure 2.9) and analytical solution proposed by Bouassida and Jellali (2002) using the generalized Prandtl's mechanism of failure (Figure 2.8).

Figures 2.10 and 2.11 clearly illustrate the closed agreement between the lower and upper numerical solutions. This demonstrates the performance of the

Table 2.2 Cases of reinforcement of soil reinforced by a trench. (With permission from ASCE.)

Initial soil	Case of reinforcement	Trench material
C = 20 kPa; γ = 17 kN/m^3 $\varphi = 0$	N°1	γ_c = 19 kN/m^3 $\varphi_c = 30°$; $C_c = K_c C$ $0 \le K_c \le 1$
	N°2	γ_c = 19 kN/m^3 $C_c = 10\,C$ = 200 kPa $0 \le \varphi_c \le 20°$

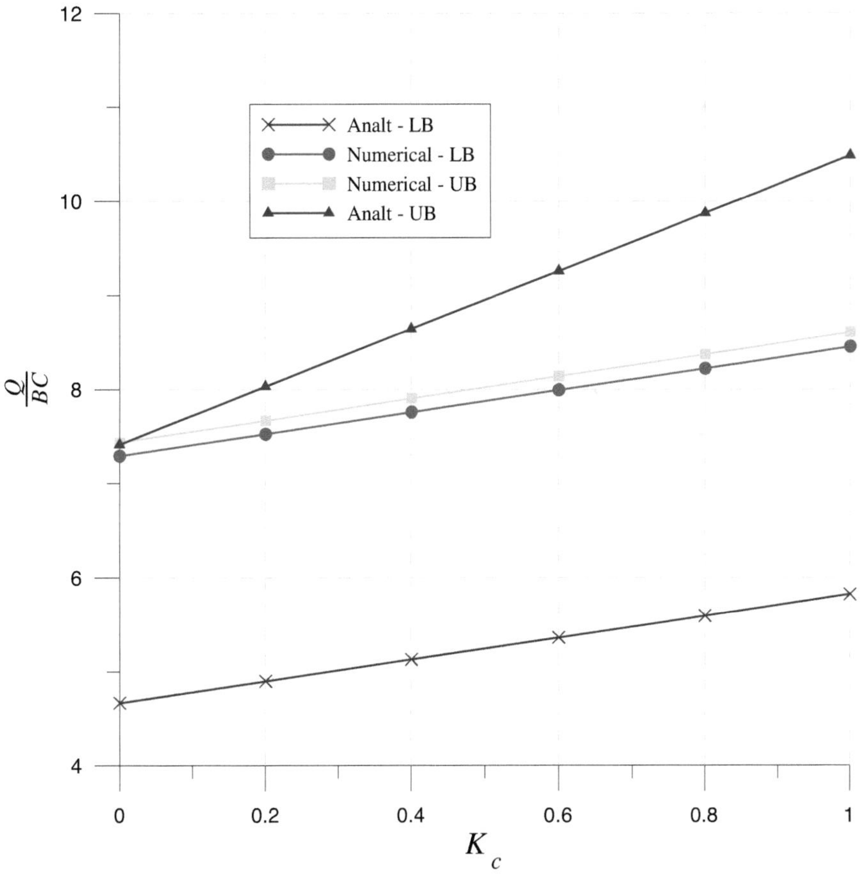

Figure 2.10 Predictions of the BCF of purely cohesive soil reinforced by a cohesive-frictional material of trench, case n°1: IAR = 1/3. (With permission from ASCE.)

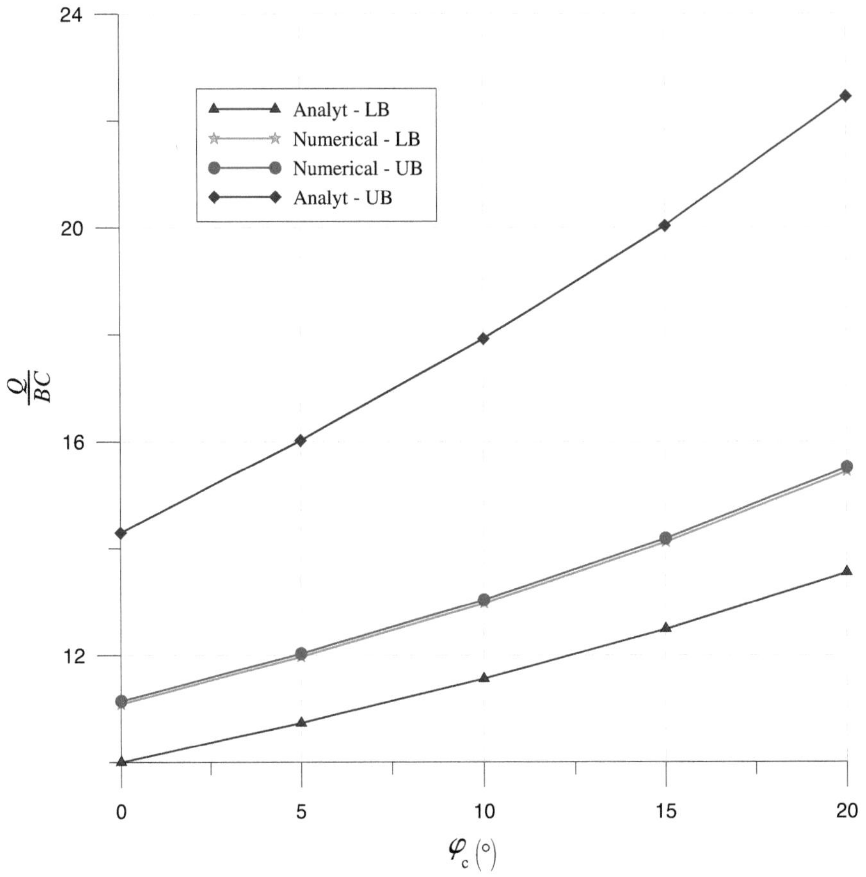

Figure 2.11 Predictions of the BCF of purely cohesive soil reinforced by a purely cohesive material of trench, case n°2: IAR = 1/3. (With permission from ASCE.)

FELA code in predicting closed solutions with a gap between numerical predictions of the ultimate bearing capacity factor less than 1.86%. It can be concluded that a quasi-exact numerical solution has been found for the bearing capacity of purely cohesive soil reinforced by a trench made up of purely cohesive or cohesive-frictional material.

> *Case of Reinforcement n°*1: the trench material is purely frictional or cohesive-frictional with a friction angle equal or less than 20 degrees.

Figure 2.10 shows that when the cohesion of a column material does not exceed half that of a purely cohesive soil, the analytical upper bound solution provided by the generalized Prandtl's failure mechanism is acceptable, since it only overestimates the quasi-exact numerical solution by 10%. In turn, the lower bound solution obtained from the overly simple five-zone stress field underestimates the numerical solution; the best approximation, 31%, is recorded when equal cohesion is considered for the two constituents of reinforced soil: $K_c = 1$.

> *Case of reinforcement n°2*: the trench material is purely cohesive or cohesive-frictional with a cohesion equal or less than that of an initial soil.

In comparison with Figure 2.10, an opposite trend is marked in Figure 2.11 that shows the analytical lower bound of the ultimate BCF predicted by the five-zone stress field better approaches the quasi-exact numerical solution than does the upper bound solution provided by the generalized Prandtl's failure mechanism. When the friction angle of trench material equals 20 degrees, the lower bound solution only underestimates the quasi exact solution by 12.25%, while the upper bound solution greatly overestimates the quasi-exact solution (at the most by 28.2% for a purely cohesive trench material).

In Figures 2.12a and b, output results by the FELA code show the plastic multipliers and failure mechanisms given by the lower bound and upper bound

(a) (b)

Figure 2.12 (a) Plastic points and power dissipation predicted by the FELA code respectively by the lower bound and upper bound solutions (reinforcement case n°1). (b) Plastic points and power dissipation predicted by the FELA code respectively by the lower bound and upper bound solutions (reinforcement case n°2). (Both with permission from ASCE.)

numerical solutions, respectively. Only half of the reinforced trench model was considered due to the geometrical and mechanical symmetries. It is noted that these numerical predictions well reproduce the mechanisms of failure shown in Figures 2.6 and 2.8 considered by the analytical contributions discussed previously.

Bouassida et al. (2015) compared the predicted plastic multipliers of cases of reinforcements between No. 1 and No. 2. They show the significant frictional effect due to the trench material obtained in case No. 1 to that shown in Figures 2.12a and b that mostly illustrate the cohesion effect of trench material. Indeed, the more trench material that is frictional, the more the discontinuity of velocity is inclined with respect to the yield surface. In turn, when the trench material is purely cohesive (case No. 2) the discontinuities of velocity coincide with the yield surface. Such observations confirm the necessary conditions on field velocities to derive the upper bound solutions by the kinematic approach of limit analysis (Salençon, 1990).

As an interpretation from the two cases of reinforcement studied, the best approximation of analytical prediction solutions are obtained when one of two constituents of reinforced soil is purely frictional and the other is mostly purely cohesive.

Bouassida et al. (2015) recently investigated the predictions of the ultimate bearing capacity of reinforced soil by a trench by performing the FLAC3D code. The FLAC three-dimensional explicit finite-difference method, incorporated in Fast Lagrangian Analysis of Continua (FLAC) code, was performed to develop the numerical computations. FLAC3D code is used for the study of the behavior of three-dimensional structures built on soils and rocks and other materials (Itasca Consulting Group, 2002).

Associated plasticity and nonassociated plasticity assumptions were adopted for the constituents of reinforced soil. Comparisons made between numerical results as obtained by the FELA and FLAC3D codes have confirmed that the quasi-exact solutions of the ultimate bearing capacity of reinforced soil by a trench are identified for four cases of study.

Although numerical results have highlighted at what extent the analytical upper and lower bounds solutions can be adopted for the prediction of ultimate bearing capacity of soil reinforced by a trench, the best confirmation of all predictions should come from an experimental investigation.

2.2.2.3 Loading Tests: Validation and Calibration of Models

Although the isolated column is a quite simple model for the prediction of bearing capacity, it has the benefit of carrying out relevant loading tests (full-scale or scaled model) to predict meaningful results. However, the isolated column model has two limitations: it does not take into account the diameter of the column, which can be arbitrary, and it ignores the group effect as to the improvement of characteristics of the initial soil due to column installation. The isolated column model has been investigated to predict the ultimate bearing capacity of the well-practiced case study of purely cohesive soft soil reinforced by stone column. Recorded data from the loading plate test are considered after Bergado et al. (1984). The data include the grain size of column material and the process of installation of 13 full scale stone column models. These field data were used for the calibration of parameters governing the elastoplastic constitutive law of a purely cohesive soft soil reinforced by a laterally expanded stone column. The ultimate bearing capacity of an isolated stone column, q_{ult}, is expressed by (Frikha et al., 2008):

$$q_{ult} = K_{pc}\left[\sigma_{h0} + c_u\left(1 + \frac{2}{1+k}Log\left(\frac{E}{3c_u(-0.1812k+0.1408)^{k-1}}\right)\right)\right] \tag{2.18}$$

σ_{h0} = horizontal at rest pressure,
$k = (1 - \sin\psi)/(1 + \sin\psi)$; ψ = angle of dilatancy of soft soil with undrained cohesion of c_u, Young modulus E, and Poisson's ratio = 0.33.

Figure 2.13 shows the variation of the predicted normalized ultimate bearing capacity of the isolated column by several models based on a lateral expanded

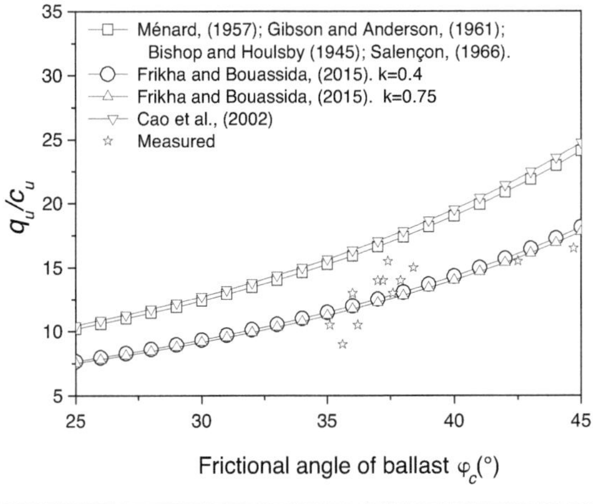

Figure 2.13 Normalized bearing capacity of an isolated column versus friction angle of stone material

cavity. It also shows the predictions by Frikha and Bouassida (2015), which match quite well with the recorded data reported by Bergado and Lam (1987). More recently, Frikha and Bouassida (2015) discussed the accuracy of various bearing capacity models for footings on single stone columns.

2.3 ULTIMATE BEARING CAPACITY OF COLUMN-REINFORCED FOUNDATION BY THE UNIT CELL MODEL

2.3.1 Previous Contributions—In Brief

Since the 1980s, the UCM has been essentially investigated for the study of experimental behavior of reinforced soil by columns with a focus on improved soft soils by granular materials. Although the estimation of settlement was essentially investigated, the prediction of ultimate bearing capacity by using the UCM was also considered. Investigations by the UCM were essentially dedicated for the determination of shear strength characteristics of reinforced soil without predicting the ultimate bearing capacity which was basically quantified by running experimental programs: Aboshi et al. (1979), Bergado et al. (1984), Bergado and Lam, (1987), and more recently Ambily and Ghandi (2007). The next section is dedicated to the study of compressive strength of a confined UCM in a limit analysis framework.

2.3.2 Equivalent and Homogenized Characteristics of Reinforced Soil

The UCM is considered, first, to estimate the equivalent characteristics of reinforced soil and to predict the bearing capacity in different ways. Second, the UCM is performed as an experimental tool for the assessment of the shear strength of reinforced soil by columns.

The UCM, depicted in Figure 2.14, was investigated by Bouassida et al. (1995) to derive the analytical compressive strength of a purely cohesive soil reinforced by cohesive-frictional column material. The geometry of the UCM is defined by its length H, the column's radius a, and the cross section A at the upper level ($z = H$) of the UCM.

The prediction of theoretical bearing capacity $(Q/A)_{th}$, in terms of the characteristics of Tunis' soft clay and reinforcing sand, for improvement area ratios $\eta = 1/25$ and $\eta = 4/25$, respectively are (Bouassida, 1996a):

$$(Q/A)_{1/25,th} = 1.144p + 20.8 \quad (2.19)$$

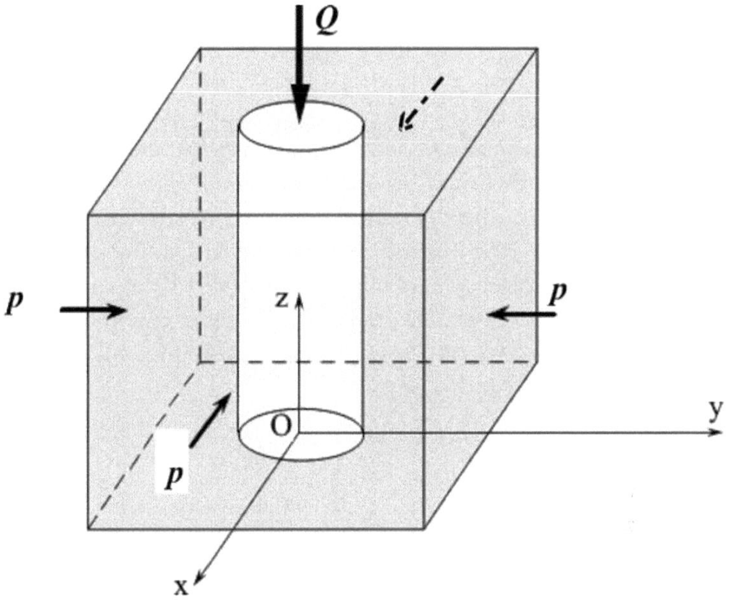

Figure 2.14 The unit cell model (Bouassida et al., 1995). Reprinted from Bouassida (1996b).

$$(Q/A)_{4/25,th} = 1.576p + 23.2 \tag{2.20}$$

Q = ultimate compressive strength that produces failure of the UCM.
p = confining horizontal pressure, expressed in kPa.

The coefficient of correlation of the linear regression in Equations (2.19) and (2.20) are, 0.999 and 0.998, respectively.

The UCM has been verified experimentally on remolded Tunisian soft clay reinforced by a sand column (Bouassida, 1996a). After recorded results from consolidated undrained triaxial tests, the experimental compressive strength $(Q/A)_{exp}$ in function of the confining pressure, p (in kPa), is written:

$$(Q/A)_{1/25,exp} = 1.195p + 14.06 \tag{2.21}$$

$$(Q/A)_{4/25,exp} = 1.554p + 15.4 \tag{2.22}$$

The coefficient of correlation of the linear regression in Equations (2.21) and (2.22) are, 0.999 and 0.998, respectively.

Comparing between theoretical predictions given by Equations (2.19) and (2.20) and experimental results given by Equations (2.21) and (2.22), it is concluded to be a very good assessment of the compressive strength of purely cohesive soil reinforced by cohesive-frictional material.

2.3.3 Assessment of Predicted Characteristics with Experimental Data

For the purpose of comparison between analytical and experimental data, the correspondence between loading conditions of the UCM and classical triaxial tests has been explained by Bouassida (1996a).

Furthermore, based on the theoretical compressive strength of a confined UCM, Bouassida et al. (1995) introduced, in triaxial loading conditions, the equivalent strength characteristics of reinforced soil which depend on the properties of initial soil and column material and the improvement area ratio. The equivalent friction angle and cohesion (φ^{eq}, C^{eq}) of the reinforced soil are identified from Equations (2.23) and (2.24):

$$K_p^{eq} = \tan^2\left(\frac{\pi}{4}+\frac{\varphi^{eq}}{2}\right) = 1+\eta(K_{pc}-1) \tag{2.23}$$

$$C^{eq} = \frac{1}{\sqrt{K_p^{eq}}}\left(C_c\eta\sqrt{K_{pc}}+C(1-\eta)\right) \tag{2.24}$$

K_p^{eq} = Rankine coefficient of passive state of stress of equivalent material.

As pointed out by Bouassida et al. (1995) the use of equivalent characteristics given by Equations (2.23) and (2.24) for the design is restricted to the case of vertically loaded structures in conformity with the triaxial conditions from which they have been introduced.

Several authors (Di Maggio, 1978; Enoki et al., 1991) have instinctively adopted the so-called homogenized strength characteristics of reinforced soil which are written:

$$tg\varphi_{\text{hom}} = \eta\, tg\varphi_c + (1-\eta)tg\varphi \tag{2.25}$$

$$C_{\text{hom}} = \eta C_c + (1-\eta)C \tag{2.26}$$

As such, the use of averaged values (φ_{hom}, C_{hom}) in terms of the improvement area ratio (IAR) easily enables us to carry out the design of column-reinforced foundations by assuming the reinforced soil as an apparent homogeneous material.

Adopting either the homogenized or the equivalent characteristics relies on assessing these predictions with experimental results. In addition, the comparison between homogenized and equivalent characteristics is worthy to carry out the calculation of the bearing capacity of reinforced soil.

Results and Interpretations

Figures 2.15, 2.16, 2.17, and 2.18 show the variations of homogenized and equivalent strength characteristics, as predicted from Equations (2.23) to (2.26), for three cases of reinforcement:

1. *Purely cohesive soil reinforced by cohesive-frictional column material*: Figures 2.15a and 2.15b show opposite trends in similar proportions, while the equivalent friction angle is overestimated compared to the homogenized one. In the range of IAR = 0.04 to 0.16, this result seems to be confirmed only for the friction angle of reinforced soil, as can be seen in Figures 2.16a and 2.16b, showing the same predictions and data recorded on samples of remolded Tunisian soft clay reinforced by compacted sand (Bouassida, 1996a). For reinforcement using the sand compaction piles, the average shear strength method was adopted by several authors (Di Maggio, 1978; Barksdale and Bachus, 1983; etc.). By this method, restricted to the improvement of purely cohesive soft soil, the averaged internal friction angle depends both on the IAR and the stress concentration factor introduced by Aboshi et al. (1979). This factor is the ratio between the induced vertical stress on the head of the sand piles to that induced at the surface of soft soil. Presentation of the average shear strength method was detailed by Bergado et al. (1996).
2. *Purely frictional soil and column material*: the reinforced material is purely frictional as well, its homogenized friction angle, as predicted

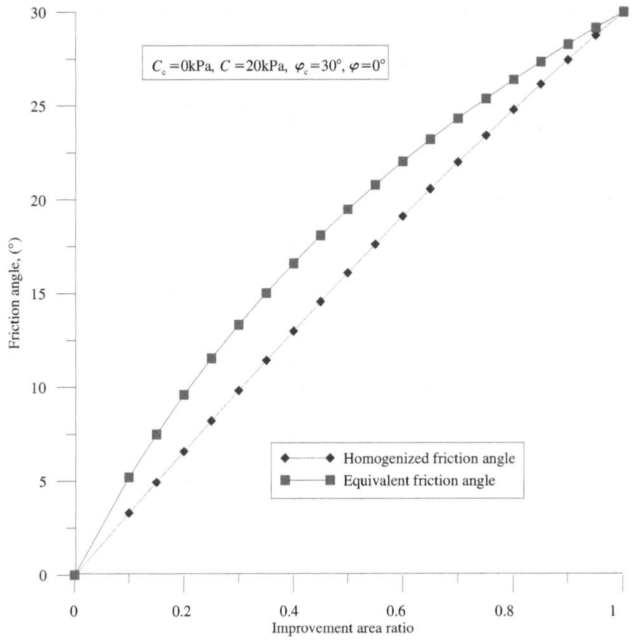

Figure 2.15a Predictions of the friction angle of purely cohesive soil reinforced by cohesive-frictional column material

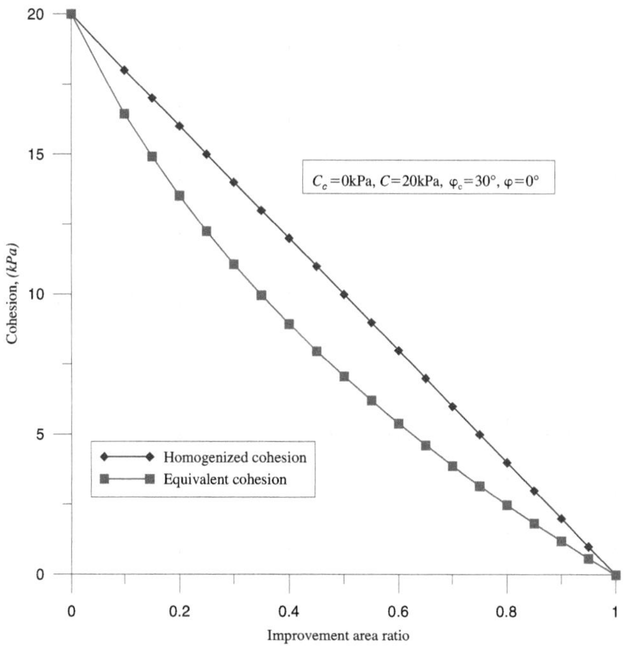

Figure 2.15b Predictions of the cohesion of purely cohesive soil reinforced by cohesive-frictional column material

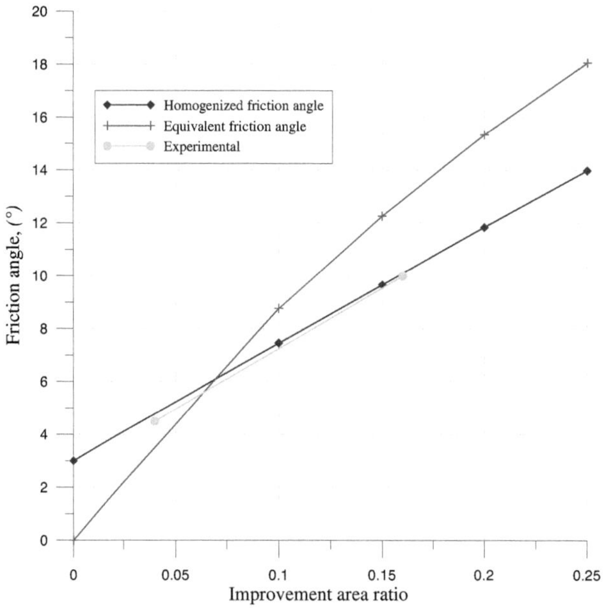

Figure 2.16a Validation of the friction angle of Tunisian soft clay reinforced by compacted sand

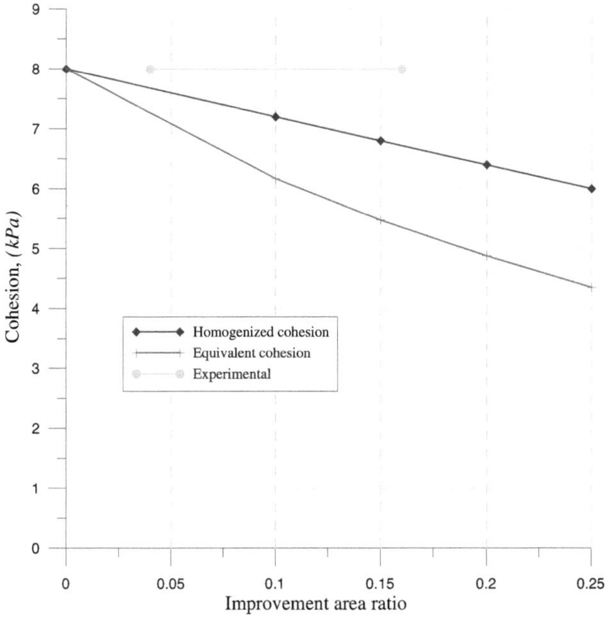

Figure 2.16b Validation of the cohesion of Tunisian soft clay reinforced by compacted sand

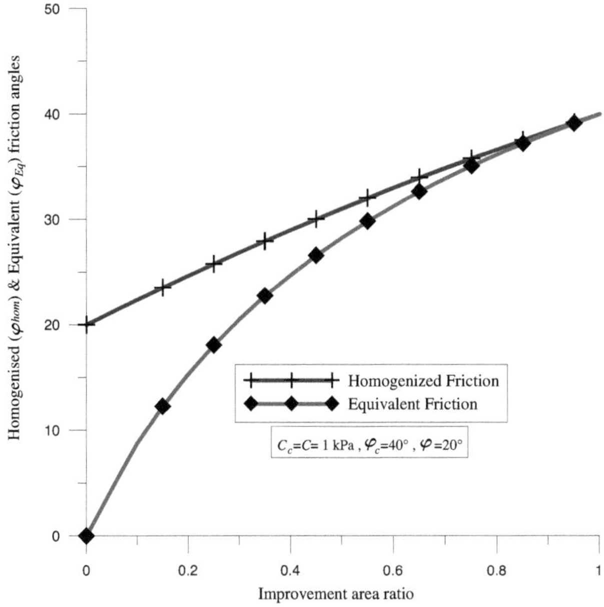

Figure 2.17 Predictions of the friction angle of purely frictional soil reinforced by purely frictional column material

by Equation (2.25), is greatly overestimated in comparison with the predicted equivalent friction angle from Equation (2.23). In the range of currently practiced IAR of 10 to 30%, this overestimation decreases from 60 to 24% (Figure 2.17).

3. *Purely cohesive soil and column material*: the reinforced soil is purely cohesive and its equivalent and homogenized cohesion have the same linear variation. In terms of dimensionless characteristics from Equation (2.26), the normalized homogenized cohesion is introduced as:

$$\langle C\rangle = C_{\mathrm{hom}}/C = 1+\eta(K_c-1) \tag{2.27}$$

Figure 2.18 illustrates the linear variation of normalized homogenized cohesion given by Equation (2.27) by using the experimental data reported by Bouassida and Porbaha (2004).

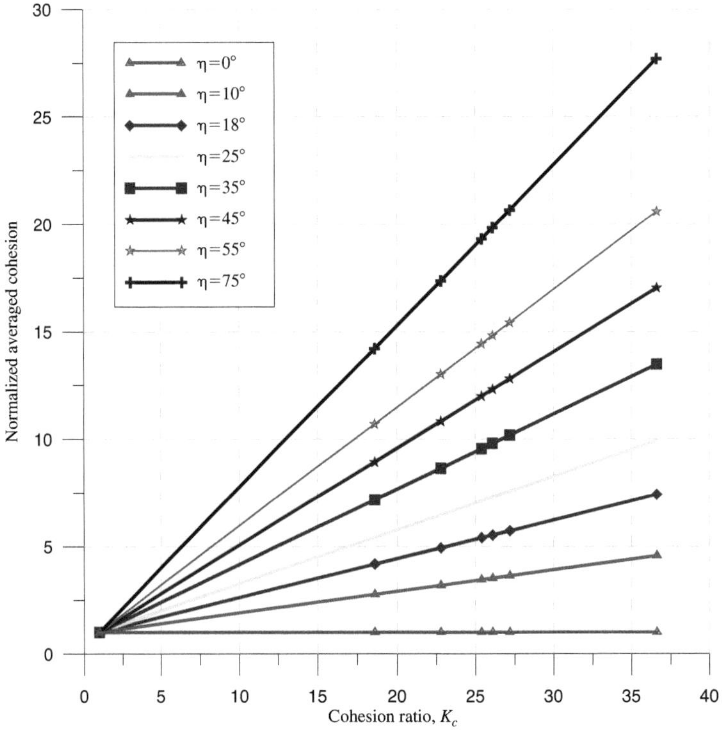

Figure 2.18 Variation of the normalized average cohesion versus the cohesion ratio and IAR using recorded data by Bouassida and Porbaha (2004)

2.4 ULTIMATE BEARING CAPACITY OF COLUMN-REINFORCED FOUNDATION BY THE GROUP OF COLUMNS MODEL

2.4.1 Direct Approaches of Limit Analysis

The modeling of reinforced soil by a group of columns was introduced in Chapter 1 (Section 1.3.3, Figures 1.7 and 1.8). The geometry of columns is defined by their length H_c and total cross section A_c of columns all located under the loaded foundation of area A.

Based on the UCM, previously detailed in Section 2.3, Bouassida et al. (1995) were the first to predict the lower bound of the BCF of a purely cohesive soil reinforced by a group of end-bearing columns made up of cohesive-frictional material. In view of predicting the best lower bound of the BCF, an optimized spacing between columns which depends on the friction angle of the column material is identified (Bouassida and Hadhri, 1998). Indeed, with the friction angle of column material and total foundation area being known, the optimized spacing corresponds to periodic adjacent intermediate areas defined by their radius b, which fulfills the condition:

$$2 \leq K_{pc} \leq 1 + (b/a)^2 \tag{2.28}$$

b = the radius of the equivalent cylindrical area, which equals that of the periodical unit cell as identified from the considered columns pattern.

The optimized IAR also depends on the type of regular grid of columns having identical radius a; it has been identified for squared and triangular grids respectively as (Bouassida and Hadhri, 1998):

$$\eta_{opt,square} = \frac{\pi}{4(K_{pc} - 1)} \tag{2.29}$$

$$\eta_{opt,triangle} = \frac{\pi}{3\sqrt{3}(K_{pc} - 1)} \tag{2.30}$$

In the case of purely cohesive soft soil reinforced by purely cohesive column material, which fairly represents the deep mixing reinforcement, a bounding of the BCF was suggested by Bouassida (1996b). It is written in terms of the normalized homogenized cohesion as follows:

$$2\left(1 + \langle C \rangle\right) \leq \frac{Q^*}{CA} \leq 2\left(1.828 + \langle C \rangle\right) \tag{2.31}$$

Inequality (2.31) provides a general solution of the ultimate bearing capacity of purely cohesive soils reinforced by purely cohesive column material as a function of the normalized homogenized cohesion given by Equation (2.27).

Two sets of recorded data were obtained from loaded scaled test models up to failure conducted on highly compressible soft clays reinforced by cement columns. These experimental data are used to validate the lower and upper bounds given by Inequality (2.31). In the range of cohesion ratio: $18 \le K_c \le 37$ and IAR = 0.18, Bouassida and Porbaha (2004) compared the analytical predictions with experimental records along with those given by the method of Broms (1982). Figure 2.19 indicates a thorough assessment of the predictions given by Inequality (2.31), which fit with experimental results. In parallel, the prediction by Broms' method seems to be conservative for a normalized homogenized cohesion greater than 10.

Mazaki et al. (2000) ran centrifuge test models on reinforced soil by a group of columns using the deep mixing method in the range of cohesion ratio $62 \le K_c \le 117$ and IAR = 79 percent. Figure 2.20 confirms that the analytical predictions obtained from Inequality (2.31) are in good agreement with the recorded data, while those given by the methods of Broms (1982) and Terashi and Tanaka (1981) significantly underestimate the experimental results.

Kasama et al. (2006) performed numerical limit analysis calculation of the ultimate bearing capacity of soil-cement columns. Numerical results provided the same upper bound of the BCF given in Inequality (2.31). Hence, for the case of reinforcement using the deep mixing technique in a wide range of cohesion

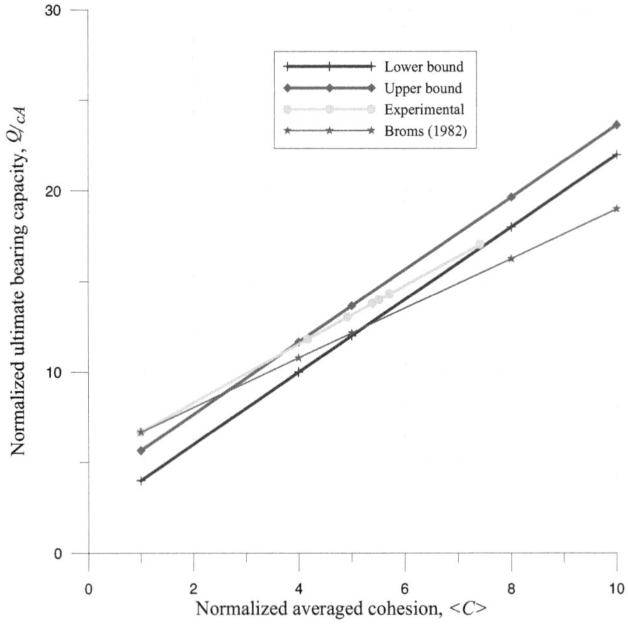

Figure 2.19 Experimental validation of the normalized ultimate bearing capacity of purely cohesive reinforced soil in the range of cohesion ratio 18 to 37

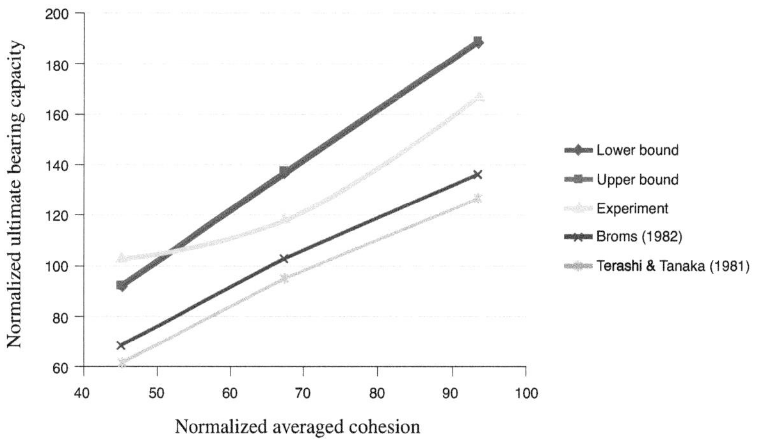

Figure 2.20 Experimental validation of the normalized ultimate bearing capacity of purely cohesive reinforced soil in the range of cohesion ratio 62 to 117

ratios and IAR, analytical predictions by the limit analysis method can be considered validated. This constitutes a more reliable tool of prediction than the previously suggested methods for the estimation of bearing capacity of reinforced soil by a group of columns. The results of bearing capacity obtained by direct approaches of limit analysis were previously discussed in Bouassida and Hadhri (1996) and Bouassida et al. (1996).

2.4.2 Floating Columns

The lower bound estimation in limit analysis framework has been extended for analyzing the reinforcement configuration by floating columns. Bouassida et al. (2009) discussed the predictions of the ultimate BCF of soil that has been reinforced by a group of floating columns by taking into account the unit weights of column material and native soil respectively denoted by γ_c and γ. The results revealed that the difference in unit weights $(\gamma_c - \gamma)$ plays an important role in predicting a maximum length of columns $H_{c,max}$ which depends on the case of reinforcement. The maximum length of columns is required after the verification of strength criterion given by Equation (2.16) for the sublayer underneath the column-reinforced soil mass (Bouassida et al., 2009). Meanwhile the BCF solutions obtained for end-bearing columns mostly remain valid in the case of floating columns. Consider, for instance, the case of reinforced soil with characteristics obeying the condition:

$$K_{pc} \geq \frac{\gamma_c}{\gamma K_p} \tag{2.32a}$$

Combining Inequality (2.32a) and the required condition on the maximum length of columns gives us:

$$K_{pc} \geq K_p \quad (2.32b)$$

Inequality (2.32b) implies that $\varphi_c \geq \varphi$, hence it applies both for the reinforcement of soft clays by granular materials and the vibro-compaction of loose sands. As illustration, consider the reinforcement of purely cohesive soil ($\varphi = 0$) by a cohesive-frictional material ($\varphi_c \succ 0$; $C_c \succ 0$). Then, the maximum length of columns is (Bouassida et al., 2009):

$$H_{\max} = \frac{2C}{\gamma_c - \gamma}\left[2 - \frac{1}{K_{pc}} + K_c \frac{1}{\sqrt{K_{pc}}}\right] \quad (2.33)$$

Therefore, the lower bound of the BCF is:

$$\frac{Q^-}{CA} = 4(1-\eta) + 2\eta(K_{pc} + K_c\sqrt{K_{pc}}) - \eta(\gamma_c - \gamma K_{pc})\frac{H_c}{c} \quad (2.34)$$

Inequality (2.33) is only valid when $\gamma_c \succ \gamma$, which is very common in practice.

From Equations (2.33) and (2.34) it can be deduced that the presence of columns is a surplus of loading, added to the load Q applied by the foundation, which contributes to the decrease of the lower bound of the BCF. For the purpose of illustration, consider the reinforcement of a very thick soft clay layer with parameters: $\gamma = 16.5 kN/m^3$, C = $15kPa$ by floating compacted sand columns of radius $a = 0.3m$ and having characteristics: $\gamma_c = 18 kN/m^3$; $C_c = 0$; and $\varphi_c = 37°$. This set of data verifies the conditions for Inequalities (2.32a) and (2.32b) and predicts a maximum length of columns $H_{c,\,min} = 35m$. In turn, by the Brauns' method from Equation (2.15), the predicted minimum length of columns is $H_{c,\,min} = 1.47m$.

Notice that the minimum length of columns predicted by the Brauns' approach gives a necessary condition that usually holds in practice since the length of columns is always over 3 m. Meanwhile, using the limit analysis method, the prediction of the maximum length of columns provides more flexibility in the way the length of column will be considered, and further, for the verification of settlement.

The ultimate bearing capacity of soft clay reinforced by a group of floating soil-cement columns subjected to a rigid foundation was investigated by Broms (1982), who adopted the mode of failure proposed for floating rigid piles (Figure 2.21). Such a failure mechanism can only be justified in the range of high IAR (IAR > 50%), which only applies for the deep mixing treatment and much enhanced cohesion of soil-cement columns. The ultimate BCF of a rigid foundation with length L and width B, is predicted as the sum of the soft clay resistance at the tip of the columns and the skin friction developed on the lateral

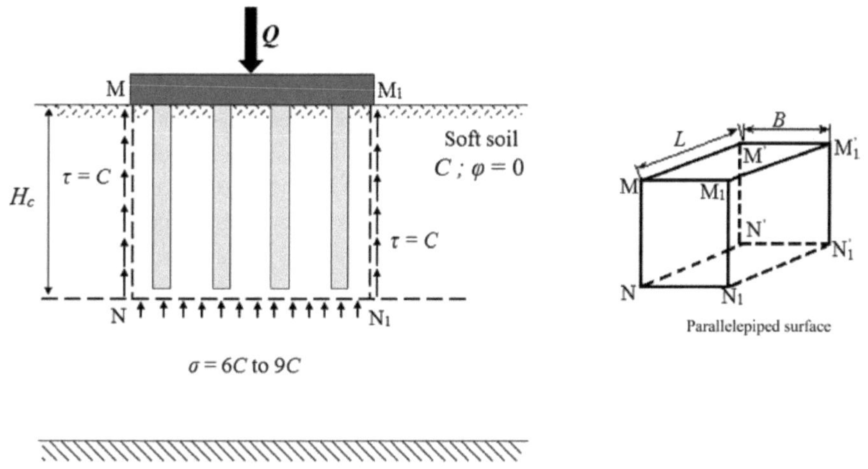

Figure 2.21 Mechanism of failure of soil reinforced by a group of floating columns (Broms, 2000). Reprinted from Bouassida (1996b).

parallelepiped area surrounding the edges of the foundation along the length of the columns. The ultimate bearing capacity is:

$$\frac{Q}{CA} = 6 + 2H_c\left(\frac{1}{B} + \frac{1}{L}\right) \tag{2.35}$$

From Equation (2.35), the prediction of the BCF does not depend on the properties of the column material and improvement area ratio, which represents the limitation of Broms' approach. Study of the case of reinforcement by end-bearing columns has shown that Broms' approach has been found conservative with respect to limit analysis predictions (Figures 2.19 and 2.20). In this regard, it can be concluded that the use of Broms' approach leads to an underestimated BCF in the case of reinforcement by floating columns. Indeed, the ultimate bearing capacity of a foundation resting on purely cohesive soft soil reinforced by purely cohesive column material can be predicted from Equation (2.34) by setting $K_p = K_{pc} = 1$. Comparing this latter prediction with Broms' prediction given by Equation (2.35), one obtains the cohesion ratio from which the BCF will be definitely underestimated by Broms' approach regarding the lower bound predicted by the limit analysis method, we come up with:

$$K_c = 1 + \frac{1}{\eta}\left(1 + \frac{H_c}{B} + \frac{H_c}{L}\right) + (\gamma_c - \gamma)\frac{H_c}{2C} \tag{2.36}$$

As an example, consider the data: $\eta = 0.16$, $H_c = 10m$, $H = 20m$; $\gamma = 16kN/m^3$; $C = 15kPa$; $\gamma_c = 17.5kN/m^3$; $B = 25m$ and $L = 35m$ to be substituted in Equation (2.36), one obtains $K_c = 12$. Then, applicability of Broms' approach remains acceptable in a very restricted range of the cohesion ratio. As a matter of fact, the relative shear strength of cement-lime treated soil to that of soft clay mostly fits in the range $K_c = 15$ to 40. This result confirms that the prediction of bearing capacity using Broms' approach is underestimated when $K_c \succ 10$ as mentioned by Bouassida and Porbaha (2004). However, this finding is not compatible with consideration of the mechanism of failure suggested by Broms, which is assumed to be developed along a parallepepidic surface within soft soil mass that encloses the reinforced soil—especially when the cohesion ratio and IAR are both high (Figure 2.21). For these reasons, Bouassida et al. (2009) evidenced that the Prandtl's failure mechanism is more efficient than the Broms' one for estimating the bearing capacity of a floating column-reinforced foundation.

2.4.3 Homogenization Method

Direct approaches of limit analysis only provided conservative lower bounds of the BCF for the reinforcement of purely cohesive soil by cohesive-frictional column material. When the reinforced soil area exceeds the area of loaded foundation, an effective lower bound solution of the BCF cannot be obtained by means of the static approach. Accordingly, the homogenization method was investigated in a second step to deepen the prediction of the bearing capacity of the CRF within the limit analysis framework, either for providing novel solutions or to improve those obtained by the direct approaches of limit analysis (Jellali et al., 2005). The homogenization solution passes, first, via the unit cell concept by the determination of the strength criterion of the homogenized medium, which actually has anisotropic characteristics. The classical approaches of limit analysis are then implemented to look for the BCF of reinforced soil (Figure 2.22). The first investigation was dedicated to reinforced purely cohesive initial soil and column material, which fits more with the deep mixing technique. Jellali et al. (2005) formulated a fairly accurate bounding of the strength criterion of homogenized medium and a somewhat better upper bound solution of the BCF of reinforced soil compared to that predicted by the direct approach of limit analysis.

The diagram (S, T) in Figure 2.23 shows the strength criterion of homogenized purely cohesive media which depends on the cohesion ratio, IAR, homogenized cohesion, and the direction defined by the inclination α that is the angle made between the major principal stress, Σ_1, with respect to the x axis. Variables $S = (\Sigma_{xx} - \Sigma_{yy})/\sqrt{2}$ and $T = \sqrt{2}\Sigma_{xy}$ are expressed in terms of the macroscopic stress field components from which the lower bound of anisotropic cohesion $c^{-}(\alpha)$ is predicted by performing the static approach of limit analysis. The upper bound of anisotropic cohesion $c^{+}(\alpha)$is predicted by performing the kinematic approach of limit analysis (Jellali et al., 2005).

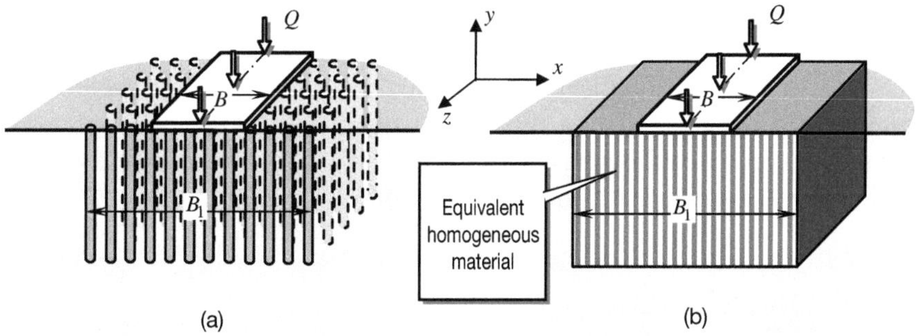

Figure 2.22 Initial (a) and associated (b) homogeneous bearing capacity problems (Jellali et al., 2005). (With permission of John Wiley & Sons, Ltd.)

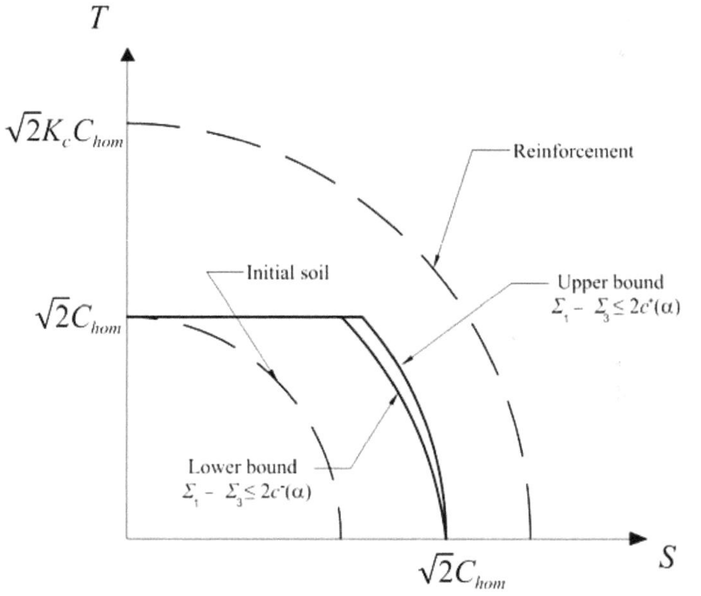

Figure 2.23 Upper and lower bound estimates of the macroscopic strength criterion of reinforced soil in plane strain condition (Jellali et al., 2005). (With permission of John Wiley & Sons, Ltd.)

A design chart has been proposed to predict the upper bound estimate of the BCF of a rigid foundation having variable width and resting on homogenized purely cohesive media (Jellali et al., 2005).

Next, Jellali et al. (2007) addressed the reinforcement of purely cohesive soft soil reinforced by cohesive-frictional column material. Although the homogenization method enabled a better prediction of the lower bound of the BCF, the determination of an upper bound solution of the strength criterion of reinforced soil is very challenging. Recently, the homogenization method has been adopted again to address the stability of an embankment resting on soft soil reinforced by end-bearing columns. Jellali et al. (2011) have proposed suitable design charts to predict the factor of stability of embankment.

2.5 CONCLUSION

This chapter covered an updated panorama on the suggested methods that deal with the prediction of the ultimate bearing capacity of a foundation resting on soils that are reinforced by columns. Classification and criticism of these methods of prediction, and their validation—either by experimental or by numerical investigations—have been discussed.

REFERENCES

Aboshi, H., Ichimoto, E., Enoki, M. and Harada, K., The "Compozer"—a method to improve characteristics of soft clays by inclusions of large diameter sand columns. Proc. Int. Symp. Reinforced soils. Paris, 1, 1979, 211–216.

Ambily, P. and Gandhi, S. R., Behavior of stone columns based on experimental and FEM analysis. *J. Geotech. Geoenviron. Eng.* 133 (4), 2007, ASCE, 405–415.

Barksdale, R. D. and Bachus, R. C., Design and construction of stone columns. Final Report FHWA/RD 83-026. U.S. Federal Highways Administration, 1983, Washington.

Barksdale, R. D., Application of the state of the art of stone columns—liquefaction, local bearing failure, and example calculations. Department of the Army, U.S. Army Corps of Engineers, Washington, D.C. 1987. Technical Report No. REMR-GT-7.

Bell, A. L., The lateral pressure and resistance of clay, and the supporting power of clay foundations. Proc. ICE 199, Part I, 1914–1915.

Bergado, D., Rantucci, G. and Widodo, S., Full scale load tests of granular piles and sand drains in soft Bangkok clay. Proc. Int. Symp. "Soil reinforcement," ENPC, Paris, 1984, 111–118.

Bergado, D. T., Anderson, L. R., Miura, N. and Balasubramaniuam, A. S., Soft Ground Improvement in Lowland and Other Environments. ASCE Press. New York, 1996.

Bergado, D. T. and Lam, F. L., Full scale load test on granular piles with different densities and different proportions of gravel and sands on soft Bangkok clay. *Soils and Foundations J.*, 27 (1), 1987, 86–93.

Bouassida, M., de Buhan, P. and Dormieux, L., Bearing capacity of a foundation resting on a soil reinforced by a group of columns. *Géotechnique*, 45 (1), 1995, 25–34.

Bouassida, M., de Buhan, P. and Dormieux, L., Discussion: Bearing capacity of a foundation resting on a soil improved by a group of columns. *Géotechnique*, 46 (3), 1996, 570–572.

Bouassida, M., Determination of bearing capacity of rigid foundations resting on reinforced soil by columns (In French). Doctorate ès-Sciences. Nat. Eng. Sch. of Tunis, 1996b, 1–246.

Bouassida, M. and Hadhri, T., Extreme load of soils reinforced by columns: The case of an isolated column. *Soils and Foundations*, 35 (1), 1995, 21–36.

Bouassida, M., Etude expérimentale du renforcement de la vase de Tunis par colonnes de sable—Application pour la validation de la résistance en compression théorique d'une cellule composite confinée. *Revue Française de Géotechnique*, 75 (2), 1996a, 3–12.

Bouassida, M. and Hadhri, T., Extreme load of soils reinforced by columns: The case of an isolated column. Closure to a discussion presented by A. Porbaha, *Soils and Foundations*, 36 (1), 1996, 118–119.

Bouassida, M. and Hadhri, T., Capacité portante d'une fondation posée sur un sol renforcé par un groupe de colonnes. *Revue Marocaine de Génie Civil*, 78, 1998, 2–16.

Bouassida, M. and Jellali, B., Capacité portante d'un sol renforcé par une tranchée. *Revue Française de Génie Civil*, 6 (7–8), 2002, 1381–1395.

Bouassida, M. and Porbaha, A., Ultimate bearing capacity of soft clays reinforced by a group of columns—Application to a deep mixing technique. *Soils and Foundations*, 44 (3), 2004, 91–101.

Bouassida, M., Jellali, B. and Porbaha, A., Limit analysis of rigid foundations on floating columns. *Int. J. of Geomech.*, ASCE. 9 (3) May-June 2009, 89–101.

Bouassida, M., Jellali, B. and Lyamin, A., Ultimate Bearing Capacity of a Strip Footing on Ground Reinforced by a Trench. Int. J. Geomech., ASCE. 15 (3), 06014021, June 2015, 1–8.

Brauns, J., Die Anfangstraglast von Schottersaûlen im bindingen Untergrund. *Die Bautechnik*, 55 (8), 1978, 263–271.

Broms, B. B., Lime and lime-cement columns: Summary and Visions. Proc. 4th Int. Conf. on Ground Improvement Geosystems, Keynote Lecture, Helsinki, 7th-9th June, 2000, 43–93.

Broms, B. B., Lime columns in theory and practice. Proc. Int. Conf. Soil Mech. Mexico, 1982, 149–165.

Datye, K. R., Settlement and bearing capacity of foundation system with stone columns. Proc. Int. Symp. Soil and rock improvement techniques including geotextiles, reinforced earth and modern piling methods. AIT–Bangkok, 1982, A1 1–27.

Di Maggio, J. M., Stone columns. A foundation treatment (In situ stabilization of cohesive soils), Demonstration project No. 4–6, Federal Highway Administration, Region 15, 1978, Arlington, Virginia, USA.

Drucker, D. C. and Prager, W., Soil mechanics and plastic analysis of limit design. *Quarterly of App. Math.*, 10, 1952, 157–165.

Enoki, M., Yagi, N., Yatabe, R. and Ichimoto, E., Shearing characteristic of composite ground and its application to stability analysis. *Deep Foundation Improvements: Design, Construction and Testing*, ASTM STP 1089, 1991, 19–31.

Frikha, W., Bouassida, M. and Canou, J., Calibration of an elastoplastic model for the prediction of stone column ultimate bearing capacity. *Proc. Geo Congress geosustainability and geohazard mitigation.* Special Pub. ASCE GEO Institute, 2008, 604–611.

Frikha, W. and Bouassida, M., Prediction of stone column ultimate bearing capacity using expansion cavity model. *Proceedings of the Institution of Civil Engineers Ground Improvement.* 168 (2), 2015, 106–115.

Greenwood, D. A., Mechanical improvement of soils below ground surface. Symp on Ground Engineering, Proc. ICE, London, paper II, 1970, 11–22.

Hughes, J. M. O. and Withers, N. J., Reinforcing of soft cohesive soils with stone columns. Proc. ICE, *Ground Engineering*, 7 (3), May 1974, 42–49.

Hughes, J. M. O., Withers, N. J. and Greenwood, D.A., A field trial of reinforcing effects of stone columns in soil. *Géotechnique*, 25 (1), 1975, 31–44.

Itasca Consulting Group. FLAC3D—Fast Lagrangian Analysis of Continua in 3 Dimensions. Version 3.00 2002 [computer program]. Itasca Consulting Group, Minneapolis, Minnesota (U.S.).

Jellali, B., Bouassida, M. and de Buhan, P., A Homogenization method for estimating the bearing capacity of soils reinforced by columns. *Int. J. of Num. & Analyt. Meth. in Geomech.* 29 (10), 2005, 989–1004.

Jellali, B., Bouassida, M. and de Buhan, P., A homogenization approach to estimate the ultimate bearing capacity of a stone column reinforced foundation. *Int. J. of Geotech. Eng.* 1, 2007, 61–69.

Jellali, B., Bouassida, M. and de Buhan, P., Stability analysis of an embankment resting upon a column-reinforced soil. *Int. J. of Num. & Analyt. Meth. in Geomech.* 35, 2011, 1243–1256.

Kasama, K., Zen, K. and Whittle, A. J., Effects of spatial variability of cement-treated soil on undrained bearing capacity. Proc. Num. Model. of Const.

Processes in Geotech. Eng. for Urban Env. Edit. T. Triantafyllidis, Ruhr Univ. Bochum, Germany, May 26th, 2006, 305–313.

Madhav, M. R. and Vitkar, P. P., Strip footing on weak clay stabilized with granular trench or pile. *Canadian Geotech. J.*, 15 (4), 1978, 605–609.

Madhav, M. R., Iyengar, N. G. R., Vitkar, P. P. and Nandia, A., Increased bearing capacity and reduced settlements due to inclusions in soil. Proc. Int. Conf. on Soil Reinforcement, Reinforced earth and other techniques, Paris, 1, 1979, 329–333.

Mazaki, K., Kimihiko, O. and Shogo, M., Centrifuge model tests on failure envelope of column type deep mixing method improved group. *Soils and Foundations*, 40 (4), 2000, 43–55.

Salençon, J., An introduction to the yield design theory and its applications to soil mechanics. *Europ. J. of Mech.*, A. Solids, 9 (5), 1990, 477–500.

Soyez, B., Méthodes de dimensionnement des colonnes ballastées. *Bulletin de Liaison des Laboratoires des Ponts et Chaussées*, 135, Janvier-Février 1985, Paris, 35–51.

Terashi, M. and Tanaka, H., Ground improved by deep mixing method, Proc. 10th Int. Conf. Soil Mech. and Found. Eng., Stockholm, 3, 1981, 777–780.

Vesic, A. S., Expansion of cavities in infinite soil mass. *J. of the Soil Mechanics and Foundation Engineering Division*, ASCE, 98, SM3, 1972, 265–290.

3

SETTLEMENT PREDICTIONS OF FOUNDATIONS ON REINFORCED SOIL BY COLUMNS

ABSTRACT

Methods for the prediction of settlement of foundations on reinforced soil (RS) by columns are introduced and then classified into two categories: empirical and rational methods. The review of rational methods first considers the modeling of RS: the unit cell model (UCM) and the group of columns model. Then, for each type of modeling, the approach used (e.g., direct or homogenization) and the constitutive laws adopted for constituents of RS by columns are considered. A method for the prediction of settlement using the group of columns model and the linear elastic behavior framework is detailed. Assessment of the prediction of settlement by several methods is discussed through comparison with recorded data related to two case histories of foundations on soils reinforced by columns.

3.1 INTRODUCTION

Engineers usually start the design of a column-reinforced foundation (CRF) by considering the reduction in settlement produced by the reinforcement.

However, as detailed in the previous chapter, the reinforcement by columns also causes an increase in bearing capacity, and quantifying this increase should be the first step in the design process. Because the reduction in settlement is a vital step in design, great importance should be placed on the study of settlement of foundations on RS by columns (Mitchell and Huber, 1985). When the column material has enhanced drainage properties, the acceleration of settlement generated by the drainage to the columns represents another potential benefit of the reinforcement.

For a preliminary design, the settlement of a foundation resting on RS should be estimated for working loads which are established from the allowable bearing capacity of the foundation. Before the 1970s, there were no rational methods available for the prediction of settlement of foundations on RS by columns. Thus, the initial designs were based on past experiences and these then formed the first principles used in the design of RS.

Some of the first rational methods for the prediction of settlement of RS by columns were published by researchers at the University of Sydney (Australia). Several developments made by Balaam, Booker, and Poulos late in the 1970s and 1980s represent milestones in the field of design of RS by columns. Since that time, these works have been followed by numerous contributions.

All suggested methods for the prediction of settlement adopted the assumption of a uniform settlement, δ, at the surface of RS (Figure 3.1). Indeed, this assumption has been confirmed from the observed behavior of vertically loaded rigid foundations (footings, rafts) on RS by columns. Reinforcing columns are basically installed under the loaded foundation as sketched in Figure 3.1. However, for some project types the installation of reinforcing columns outside the loaded foundation area is recommended, e.g., the API standard for tanks projects.

Several contributions have been suggested for the design of foundations on soils reinforced by columns on the basis of settlement criteria. The UCM was mostly employed because columns are often installed in a regular triangular or square pattern. When this UCM is used and the columns are taken to be end-bearing, the model is axisymmetric, which greatly simplifies the analysis. Additionally, model test experiments can be performed in oedometer conditions. A multitude of contributions has been reported on the UCM as a tool for investigating the behavior of soils reinforced by columns. Analytical solutions and experimental investigations have made possible the quantification of bearing capacity and settlement predictions; e.g., Ghionna and Jamiliowski (1981), Balaam and Booker (1981), Bouassida (1996), Ambily and Ghandi (2007).

Nevertheless, when studying the reinforcement by columns installed in an arbitrary pattern under a non-symmetric shaped foundation or non-uniform loading, the UCM will be inadequate. Accordingly, group of columns modeling is more realistic and preferable. Two earlier investigations for the prediction of

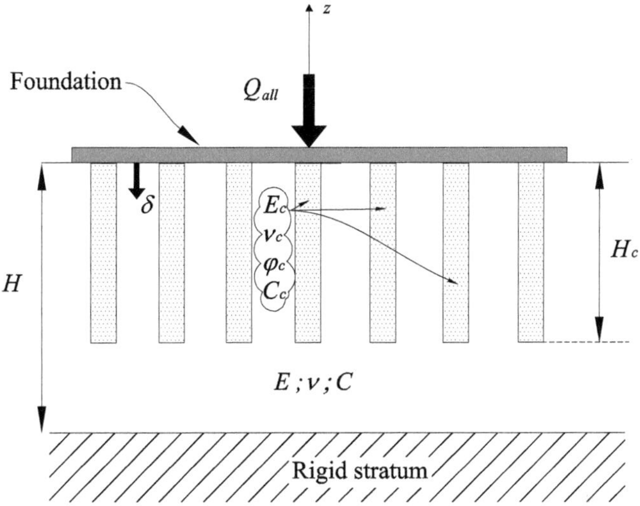

Figure 3.1 Assumed uniform settlement of a rigid foundation on reinforced soil by columns

settlement using group of columns modeling had been proposed by assuming linear elastic behavior of the constituents of reinforced soil (Broms, 1982 and 2000, Rao and Ranjan, 1985).

The prediction of settlement was essentially investigated by assuming the linear elastic behavior for the constituents of RS (Balaam and Booker, 1981). The linear elastic behavior was adopted for the prediction of settlement because of its relative simplicity. The elastic properties used in the analysis are usually derived from data recorded during triaxial tests or in situ tests. However, several investigators used elastoplastic constitutive laws, in place of linear elasticity, to model either the initial soil or the column material, e.g., Hughes et al. (1975), Datye (1982), Poorooshasb and Meyerhof (1997).

Predicted settlement by the linear elastic behavior mostly refers to the deformation of the CRF at the final stage of loading (or end of construction). At this stage, the observed settlement does not necessarily correspond to the short-term behavior of the RS, especially when the columns are used to reinforce a saturated initial soil like soft clay, and the column material is highly permeable. The columns act like vertical drains and accelerate the consolidation and, therefore, in most situations the observed settlement at the end of construction is also the long-term settlement especially for the case of end-bearing columns, which maximize the drainage for consolidation. Therefore, in practice, the long term-settlement of the CRF refers to the induced deformation within unreinforced compressible sublayers which correspond to the reinforcement by floating columns of thick compressible weak soils.

The reduction of settlement was introduced by adopting the one-dimensional modeling of RS and by assuming the applied load as an averaged uniform vertical pressure p. Because of the difference in stiffness between the column material and the initial soil ($E_c \succ E$) the factor of concentration of vertical stress has been introduced as:

$$n = \frac{\sigma_{v,c}}{\sigma_{v,s}} \tag{3.1}$$

$\sigma_{v,c}$ and $\sigma_{v,s}$ denote the vertical stresses acting at the head of columns and at the surface of the surrounding initial soil, respectively.

The reduction of settlement is defined by the factor:

$$\beta = \frac{\delta_{ur}(p)}{\delta_r(p)} \tag{3.2}$$

Where

$\delta_{ur}(p)$ = settlement of the unreinforced soil

$\delta_r(p)$ = settlement of the reinforced soil

Those settlements are calculated for the same averaged applied vertical pressure p.

Because the settlements at the top of the columns and the surface of the initial soil are assumed to be equal, the factor of stress concentration in Equation (3.1) reduces to the ratio of Young's moduli of column material to that of the initial soil. Substituting the settlements in Equation (3.2) according to the one-dimensional linear elastic behavior, the settlement reduction factor is:

$$\beta = 1 + \eta(n-1) \tag{3.3}$$

Equation (3.3) shows the more the improvement area ratio (IAR) and/or the difference in rigidities is, the more the settlement is reduced. The estimation of the concentration factor n represents the key to the problem of estimating the factor of reduction of settlement.

Soyez (1985) had reported that the determination of the factor of stress concentration n for a given IAR was essentially investigated by considering either the linear elastic behavior or the elastoplastic one. Values of n are normally in the range of 2 to 9 (Bergado et al. 1996).

This chapter mainly focuses on the methods of settlement prediction without investigating the post behavior of the improved initial soil due to either the method of columns installation or the consolidation induced by the acceleration of settlement. First, the methods for predicting the settlement of RS by columns are classified. Then, the predictions made by the UCM are reviewed, discussed, and validated. The group of columns model is then described and a method for the prediction of settlement, Bouassida et al. (2003a), which adopts the linear elastic behavior for constituents of the RS is detailed.

Finally, at the end of this chapter, an assessment is made of the settlement predictions by unit cell and group of columns modeling using data from two case studies.

3.2 CLASSIFICATION OF METHODS FOR THE PREDICTION OF SETTLEMENT OF REINFORCED SOIL BY COLUMNS

The prediction of settlement of RS by columns was reviewed by Soyez (1985), who classified two types of methods that were developed up to the early 1980s. The majority of these methods had considered the UCM and the linear elastic or elastoplastic behavior for constituents of the RS. Fessi-Guetif (2004) presented an updated review including contributions which had adopted group of columns modeling to estimate the settlement of a CRF.

In an attempt to describe the majority of published contributions in regard to the estimation of settlement of RS by columns, three main classes of methods should be distinguished (Table 3.1).

The first class comprises the empirical and semi-empirical contributions—and there are only a few of these. The second class covers all rational methods

Table 3.1 Classification of methods of settlement prediction

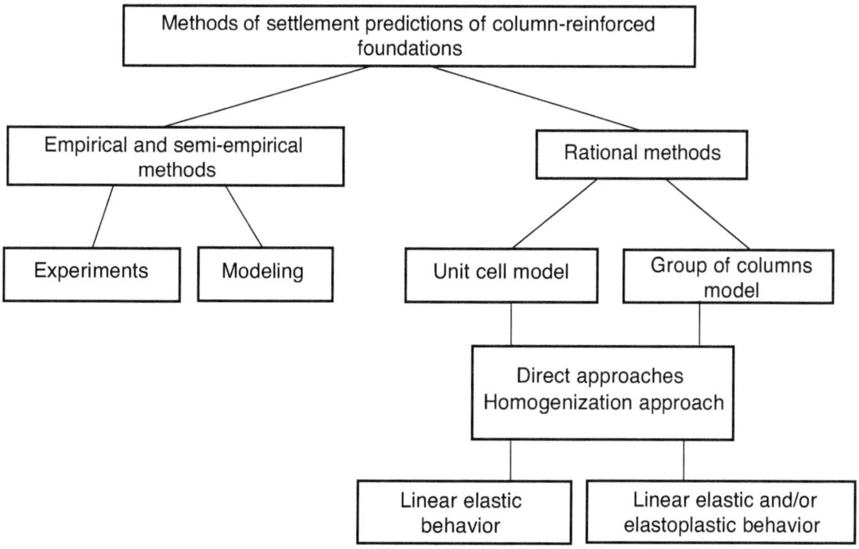

based on the use of the UCM. The third class covers all rational methods based on the use of the group of columns model. The second and third classes are subclassified into two categories. The first category of rational methods includes those using so-called direct approaches, which consider that the reinforced ground is composed of two constituents having different parameters—e.g., Young's moduli are E and E_c for the initial soil and column material. The second category of rational methods considers the homogenization approach, which assumes the RS is a mono-phase material having homogenized or equivalent characteristics described by functions of the parameters of constituents of the RS used by the direct approach. The direct and homogenization methods and the different constitutive behaviors can be adopted in the unit cell and group of columns models.

Tables 3.2 and 3.3 summarize the recommended rational methods for the prediction of settlement of foundations on RS by columns using the direct

Table 3.2 Direct approaches used for the prediction of settlement of foundations on RS by columns

Methods	Modeling of RS	Constitutive law(s)	Solution	Settlement depends on
Greenwood (1970)	None	None	Empirical	Methods of columns installation and undrained shear of host soil
Priebe (1976, 1995)	Two-dimensional Axisymmetry (UCM)	Linear elasticity	Semi-empirical	Elastic properties and friction angle of column
Chow (1996)	One-dimensional	Linear elasticity	Analytical	Averaged oedometer moduli
Balaam and Booker (1981)	Two-dimensional Axisymmetry (UCM)	Linear elasticity	Analytical	Young's moduli, Poisson's ratios, and IAR
Balaam and Poulos (1983) Balaam and Booker (1985)	Two-dimensional Axisymmetry (UCM)	Elastoplastic	Numerical code (FEM)	Elastic properties and friction angle of column
Van Impe and de Beer (1983)	Plane strain Equivalent trench	Linear elasticity	Analytical	Elastic properties and angle of dilatancy
Bouassida et al. (2003)	Three-dimensional	Linear elasticity	Analytical	Young's moduli, Poisson's ratios, and IAR

Table 3.3 Homogenization approaches used for the prediction of settlement of foundations on RS by columns

Methods	Modeling of RS	Constitutive law(s)	Solution	Settlement depends on
Schweigher and Pande (1986)	None	Weighted average properties	Analytical and numerical	Methods of columns installation and undrained shear of host soil
Canetta and Nova (1989)	Weighted average	Weighted average properties	Analytical and numerical	Anisotropic properties
Lee and Pande (1998)	One-dimensional	Linear elasticity	Analytical	Averaged oedometer moduli
Ben Said et al. (2004)	Two-dimensional Axisymmetry (UCM)	Linear elasticity	Analytical	Young's moduli, Poisson's ratios, and IAR
Abdelkrim and de Buhan (2007)	Two-dimensional Axisymmetry (UCM)	Elastoplastic	Load-settlement curve	Homogenized properties of equivalent material
Ghazi et al. (2009)	Two-dimensional Axisymmetry (UCM)	Elastoplastic	Numerical	Homogenized properties of equivalent material

approaches and the homogenization methods, respectively. It is worth noting that the majority of recommended methods have been built on the basis of the UCM and, basically, have been more or less validated through laboratory and/or experimental investigations.

3.3 EMPIRICAL AND SEMI-EMPIRICAL METHODS

3.3.1 Greenwood's Chart

Greenwood (1970) suggested the first chart for the design of rafted foundations on soft clays reinforced by end-bearing stone columns. Settlement records from case histories of stone columns installed by the vibro-substitution and vibro-displacement methods were used to correlate the reduction in long-term settlement with the spacing between columns and the undrained shear strength of soft clay (Figure 3.2). Interpretation of Greenwood's chart illustrates the following statements.

Irrespective of the tested spacing between columns in soft clay with undrained cohesion $C_u = 40$ kPa, the reduction of settlement is more significant by

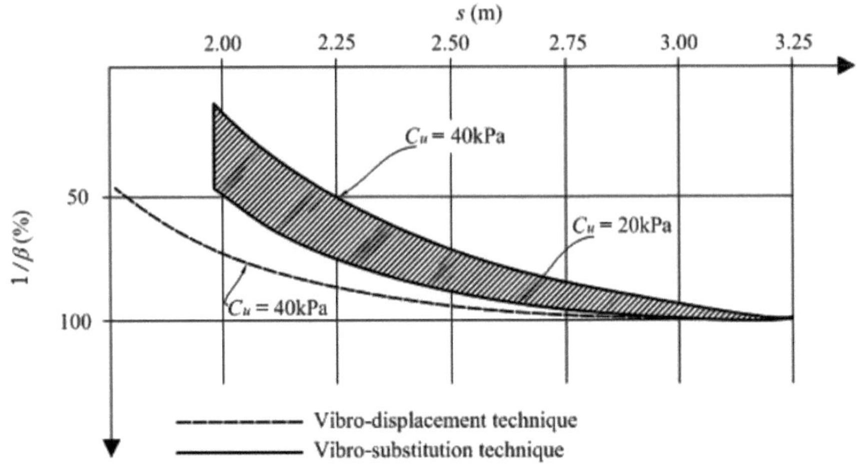

Figure 3.2 Reduction in settlement of rafted foundations on homogeneous soft clay (immediate settlement is neglected) reinforced by end-bearing stone columns (Greenwood, 1970) (with permission from the Institution of Civil Engineers)

using the vibro-substitution technique of columns installation rather than using the vibro-displacement technique. Irrespective of the technique of installation, the reduction of settlement will not be effective when the spacing between stone columns is beyond 3.25 m.

Note that for a given pattern and diameter of columns, the IAR is deduced from the spacing between columns. Balaam and Booker (1985) performed an elastoplastic finite element analysis for the study of the behavior of a stone CRF. From the numerical predictions it was found that the effectiveness of spacing between columns is more relevant than the empirical suggestion made by Greenwood (1970).

3.3.2 Priebe's Method

The basis of Priebe's method was developed some 40 years ago and then published (Priebe, 1976). Subsequently, several adaptations, extensions, and supplements have been incorporated and these modifications justify a new and comprehensive description of the method, which refers to the improvement effect of stone columns in an unaltered soil compared to the initial soil (Priebe, 1995). Priebe's method is widely used in the designs of stone CRFs that are focused on the prediction of settlement. Priebe's method was programmed in the software *Dc-Vibro* (www.dc-software.com) and later in the software *Greta* for the analysis of vibro-compacted stone CRFs (Priebe, 2003).

Many authors have presented Priebe's method and interpreted its prediction of settlement. The criticisms made by Ellouze et al. (2010) about Priebe's method can be summarized as follows:

1. In the first step an improvement factor, denoted by n_0, is introduced in which stone columns improve the mechanical characteristics of the initial soil in comparison to its pretreatment properties (without columns). According to this improvement factor, Young's modulus of the composite system is increased and, subsequently, the settlement is reduced. All further steps of design refer to this basic principle. The assumption made on factor n_0 is unclear, and no explanation is provided to derive the expression of another improvement factor n_1. Further, a depth factor depending on the unit weight of the initial soil is introduced. This is inconsistent with an assumption used in the derivation that unit weights of the RS constituents are initially neglected. Then, the incorporation of the depth factor f_d, which greatly improves the reduction of settlement needs to be clarified.
2. The reinforced ground, with a prescribed grid of stone columns, is assumed unlimited and subjected to a uniform vertical surcharge. Idealized assumptions are: the UCM is adopted; an end-bearing column, made up of incompressible material is considered; and unit weights of the initial soil and column material are neglected.
3. Some inconsistencies exist because of the assumptions made for the UCM. In the first step of the analysis, Priebe considers a cylindrical cavity subjected to lateral expansion by which zero vertical deformation is assumed that produces a solution derived from the plane strain condition. This solution is then incorporated in the UCM for which there is a distribution of vertical stress generating non-null vertical deformation, and consequently, the settlement is assumed constant. Priebe's method is carried out by using the same UCM for two different problems (from the loading condition point of view): first, the lateral expansion of a cylindrical cavity, and second, a vertical loading. Under these conditions, the use of the principle of superposition on these two states of stresses is not straightforward (Salençon, 1988).

In addition, it should be noted that Dhouib and Blondeau (2005) mentioned that the design formulas of Priebe's method appear to be of complex use due to their semi-empirical character. Later on, Dhouib (2006) confirmed this statement.

In conclusion, the steps of the design procedure are not connected mathematically and they use different simplifications and approximations.

3.4 SETTLEMENT PREDICTION BY THE UNIT CELL MODEL

The UCM, shown in Figure 3.3, is a single circular cross section column with radius a centered in a cylindrical volume of soil with external radius b and height H_c. The IAR is then given by: $\eta = a^2 / b^2$.

Contributions based on the UCM cover analytical and numerical methods of predictions by assuming an elastoplastic behavior for the column material and or for the initial soil. Among numerous works, one can cite the contributions made by Goughnour and Bayuk (1979), Balaam and Booker (1985), and Poorooshasb and Meyerhof (1997). In this section a review of the various approaches for the settlement prediction is presented. The review first considers the methods based on linear elastic behavior, then elastoplastic behavior for constituents of the RS.

3.4.1 Methods Adopting Linear Elastic Behavior

Balaam and Booker (1981)

In this method of analysis, the UCM in oedometer conditions (Figure 3.3) is used, i.e., zero horizontal displacement at the cell border ($r = b$; $0 \le z \le H_c$).

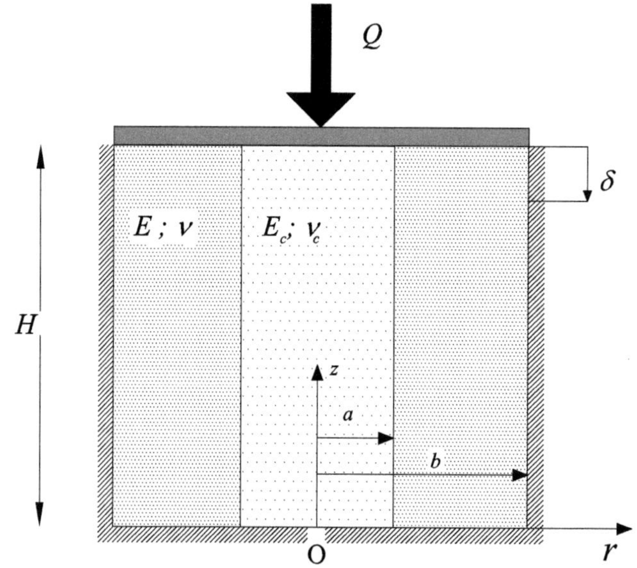

Figure 3.3 Unit cell model ($H = H_c$)

Adopting the linear elastic behavior for both the initial soil and column material, the apparent Young's modulus of the UCM was derived analytically:

$$E_{BB} = \langle E_{oed} \rangle - \frac{\eta(1-\eta)(\lambda_c - \lambda)^2}{(\lambda_c + \mu_c + \mu) + \eta(\lambda + \mu - \lambda_c - \mu_c)} \tag{3.4}$$

λ_c; μ_c and λ; μ are respectively the Lamé's coefficients of the column material and initial soil, which are related to the oedometer moduli by:

$$E_{oed(c)} = \lambda_{(c)} + 2\mu_{(c)} \tag{3.5a}$$

The oedometer modulus can be also expressed as a function of Young's modulus and Poisson's ratio as follows:

$$E_{oed(c)} = \frac{E_{(c)}(1-\nu_{(c)})}{(1-2\nu_{(c)})(1+\nu_{(c)})} \tag{3.5b}$$

$\langle E_{oed} \rangle$ is the averaged oedometer modulus, written as:

$$\langle E_{oed} \rangle = \eta E_{oed(c)} + (1-\eta) E_{oed} \tag{3.6}$$

Chow (1996)

This method was developed from Balaam and Booker's (1981) analysis by assuming zero horizontal displacement everywhere in the reinforced soil. Consequently, the expression of the apparent Young's modulus reduces to the averaged oedometer modulus expressed by Equation (3.6). The assumed zero horizontal displacement results in an upper bound value of Young's modulus of RS established by Balaam and Booker (1981). Hence, it results in a nonconservative prediction of the settlement of RS, particularly when the IAR is less than 16 percent (Chow, 1996). Therefore, the use of Chow's analysis is only recommended when $\eta \geq 0.16$. The associated prediction of settlement will be more realistic when the dimensions of loaded foundation are significantly greater than the thickness of the RS layer(s); as such, the oedometer condition prevails.

3.4.2 Methods Adopting Elastoplastic Behavior

Ghionna and Jamiolkowski (1981) considered a unit cell in oedometer conditions, a linearly elastic initial soil, and elastoplastic column material. It was assumed the loading induced a constant vertical stress in depth. By this method, the increase in horizontal stress within the column was predicted as a function of the drained elastic properties of the initial soil and the IAR.

Balaam and Poulos (1983) is more likely the first contribution that analyzed the load-settlement response of stone column-reinforced ground by implementing a finite element code and adopting the elastoplastic behavior. Using data from a full-scale load test, the suggested analysis was able to accurately reproduce the response of the foundation resting on the stone column-reinforced soil in perfect contact with the surrounding clay. The settlement reduction predictions made were in fair agreement with Priebe's method. Further, the analysis was extended for the prediction of the evolution of settlement of stone column-reinforced clay subjected to flexible and rigid foundations.

Balaam and Booker (1985) presented an interaction analysis that takes account of the yielding within the stone column by assuming it to be an elastic perfect plastic material. Accordingly, an elastoplastic finite element (FE) procedure was implemented to check the validity of assumptions linked to the behavior of constituents of the RS. The load settlement response of a rigid raft on a stabilized stone column-reinforced foundation predicted by the interaction analysis revealed good agreement with the predictions made by the FE analysis. Further, predictions made by the linear elastic behavior were found to be nonconservative in comparison to predictions made by the interaction and FE methods.

Several contributions have considered the coefficient K_0 of pressure at rest to analyze the evolution of horizontal stress within the column in pre- and post-treatment of the initial soil by stone column. Six et al. (2012) have reviewed and compared the predictions produced by three well-known contributions Priebe (1995), Goughnour (1983), and Ghionna and Jamiolkowski (1981).

Six et al. (2012) proposed a modified version of the method suggested by Poorooshasb and Meyerhof (1997) by adopting a Mohr Coulomb non-associated constitutive law for the column material. This method takes account of a horizontal residual stress to simulate the effects of stone column installation and gravity as well. This procedure has been implemented in a FE code, which was used to produce a parametric study of the effect on settlement reduction of the coefficient of the at-rest state K_0. The main finding was that the numerical prediction made by this modified elastoplastic model confirmed the analytical results in the sense that a higher value of K_0 affects the performance of stone column installation, and this is associated with the progress of plasticity within the reinforcing column.

COLANY is a computer program that computes the Balaam and Booker (1981, 1985) analyses. An analytic solution for the settlement assuming no yield in the clay or column is calculated. The program then computes the load-settlement response of the stabilized clay. Finally, the solution is computed for the settlement response of the stabilized clay when the column is fully plastic. These solutions enable the load-settlement response of a rigid raft supported by a layer of clay stabilized with a large number of stone columns to be predicted for all load levels. The validity of the assumptions made in this analysis has been checked by performing elastoplastic FE analyses, and the agreement is very good (Balaam and Booker, 1987).

3.5 SETTLEMENT PREDICTIONS BY THE GROUP OF COLUMNS MODEL

Few contributions have focused on the prediction of the settlement of a column-reinforced ground when it is modeled as a truly 3-D medium. A review of the literature indicates that Broms (1982) published the first method dealing with the prediction of settlement of RS by a group of columns made up of lime-cement-treated soft clay. The method proposed by Broms had assumed that the apparent Young's modulus of reinforced soil can be predicted by using the law of mixtures:

$$E_{app}^{0} = (1-\eta)E + \eta E_c \tag{3.7}$$

Estimations by the law of mixtures model or the weighted method have since been used by Rao and Ranjan (1995) and other authors without providing any justification of this estimate of the apparent Young's modulus of RS.

The settlement prediction of an RS by a group of columns should be addressed by a rational method using linear elastic behavior. Indeed, because of the composite nature of RS it is not possible to derive an analytical solution for the linear elastic settlement. The use of numerical methods also poses various challenges for the two- or three-dimensional modeling, i.e., the interface between the initial soil and columns that have disproportioned rigidities; the choice of constitutive laws requires the selection of adequate soil parameters, etc. For these reasons, considering energy theorems in linear elasticity, especially based on the minimum stress principle, Salençon (1988) provides an adequate alternative for estimating the settlement of RS by a group of columns.

Bouassida et al. (2003a) have presented a variational method in linear elasticity by considering the group of columns model. This implementation is used to predict a lower bound of the apparent Young's modulus of reinforced soil and this is then associated to an upper bound estimate of the settlement. This method provides a rigorous solution for the conservative settlement estimation of the CRF.

For a rigid raft foundation having a prescribed area and subjected to an allowable vertical load Q_{all}, the apparent Young's modulus of the RS is given by:

$$E_{app} = \frac{(Q_{all} / A)}{\delta / H_c} \tag{3.8}$$

Consider a circular rigid raft foundation with radius R resting on soil reinforced by a group of end-bearing columns (Figure 3.4) that are installed in an arbitrary

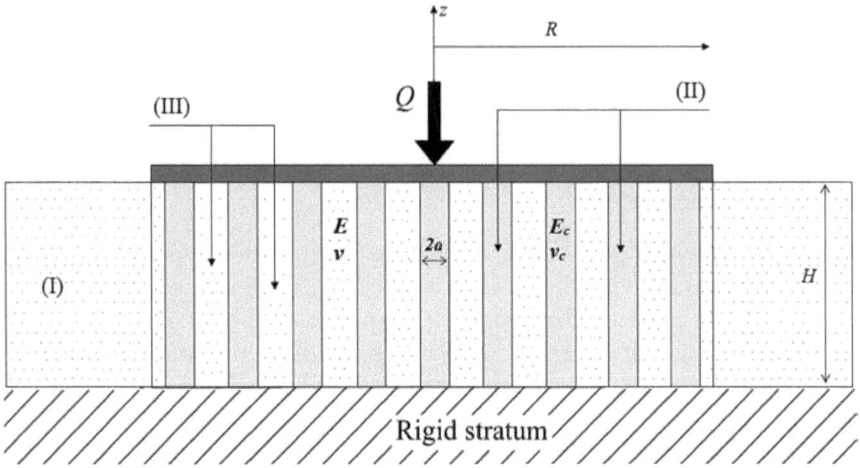

Figure 3.4 Circular rigid raft on a column-reinforced ground: three-zone stress field ($H = H_c$)

location under the foundation. At the surface of the reinforced ground, the raft foundation is subjected to a prescribed settlement $\delta > 0$, due to the loading. The linear elastic behavior of the columns (E_c; v_c) and initial soil (E; v) is considered, assuming both materials are homogeneous, isotropic, and weightless. When different unit weights for the initial soil and column material are considered, Bouassida et al. (2003a) demonstrated that the solution of the stated problem does not depend on gravity. This elastic solution is derived by establishing a statically admissible (s.a.) stress field denoted $\underline{\underline{\sigma}}$, i.e., verifying equilibrium equations, the boundary conditions, and the continuity of the stress vector between prescribed zones of the established stress field. The solution is obtained from the distribution of vertical stress, σ_z, over the area of the rigid raft foundation which balances the vertical allowable load Q_{all}, that is:

$$Q_{all} = -\int_{z=0,r\leq R} \sigma_z \, dA \tag{3.9}$$

In this section, the normal stresses are taken to be negative and therefore the compressive vertical stress, σ_z, is associated with a positive compressive load Q_{all} in Equation (3.9).

Hereafter a variational stress approach is performed to predict a lower bound of the apparent Young's modulus of column-reinforced ground modeled as a 3-D medium.

3.5.1 Variational Stress Approach in Linear Elasticity

The principle of the variational method is that the stress field of the exact solution of the stated problem minimizes, over all the s.a. stress fields, the functionality of complementary energy expressed by (Salençon, 1988):

$$U(\underline{\underline{\sigma}}')=W(\underline{\underline{\sigma}}')-\Phi(\underline{\underline{\sigma}}') \tag{3.10}$$

$\underline{\underline{\sigma}}'$ denotes any exhibited s.a. stress field. In Equation (3.10) the elastic stress energy is expressed by:

$$W(\underline{\underline{\sigma}}')=\int_{\Omega}\left[\frac{1+\nu}{E}tr\underline{\underline{\sigma}}'^2-\frac{\nu}{2E}\left(tr\underline{\underline{\sigma}}'\right)^2\right]d\Omega+\int_{\Omega_C}\left[\frac{1+\nu_C}{E_C}tr\underline{\underline{\sigma}}'^2-\frac{\nu_C}{2E_C}\left(tr\underline{\underline{\sigma}}'\right)^2\right]d\Omega_C \tag{3.11}$$

Ω and Ω_c respectively denote the domain of the initial soil and the reinforcing columns. The second functional energy in Equation (3.10) is given by:

$$\Phi(\underline{\underline{\sigma}}')=\sum_k \xi_k^g\sigma'_{ki}n_i dA_k=\int_{z=0\&r\leq R}-\delta\sigma'_{zz}\,dA \tag{3.12}$$

A_k denotes the area of the boundary of the reinforced ground system where the k^{ith} component of displacement is a given data $\xi_k=\xi_k^g$.

By using Clapeyron's theorem, the elastic stress energy for the field of stress $\underline{\underline{\sigma}}$ solution of the problem gives (gravity is neglected):

$$W(\underline{\underline{\sigma}})=\frac{1}{2}\int_{\partial\Omega}\underline{\xi}.\underline{\underline{\sigma}}.\underline{n}dS=\frac{1}{2}Q\delta=\frac{1}{2}\Phi(\underline{\underline{\sigma}}) \tag{3.13}$$

Then, from Equation (3.10) the principle of minimum complementary energy can be expressed as:

$$\forall\underline{\underline{\sigma}}'\text{s.a.}\qquad U^*(\underline{\underline{\sigma}})=-\frac{1}{2}Q\delta\leq U^*(\underline{\underline{\sigma}}') \tag{3.14}$$

Substituting Equations (3.8), (3.13), and (3.14) into Equation (3.10) one obtains:

$$\forall\underline{\underline{\sigma}}'\text{s.a.}\qquad E_{app}\geq-\frac{2H_c}{\pi R^2\delta^2}U^*(\underline{\underline{\sigma}}')\text{s.a} \tag{3.15}$$

Consider the three-zone stress field sketched in Figure 3.4, and detailed in Appendix 5.1. Then, applying the principle of minimum energy of stress leads to the lower bound estimate of the apparent Young's modulus of RS, E_{cir}^{var}, given by (Bouassida et al. 2003a):

$$E_{app}\geq E_{cir}^{var}=(1-\eta)E+\eta E_c+2(\nu^*)^2E^* \tag{3.16}$$

$$\nu^*=(1-\eta)\nu+\eta\nu_c$$

$$E^*=\left[\frac{1+\nu}{E}+\eta\frac{(1+\nu_c)(1-2\nu_c)}{E_c}+(1-\eta)\frac{(1+\nu)(1-2\nu)}{E}\right]^{-1}$$

3.5.2 Rectangular Raft on Column Reinforced Foundation

The solution in Equation (3.16) for a rigid circular raft can be extended for a rectangular foundation of length L and width B. Indeed, as shown in Figure 3.5, consider the three-zone stress field presented in Figure 3.4, for which a fourth zone is added between the circumscribed circle with radius $R=\sqrt{B^2+L^2}$ and the rectangular foundation. By implementing the same procedure carried out for the case of a circular foundation, the case of a rectangular foundation is solved. The minimization over the parameters of the constructed four-zone stress field leads to the following lower bound solution of the apparent Young's modulus of RS (Bouassida et al. 2003a):

$$E_{app} \geq E_{rec}^{var} = (1-\eta)E + \eta E_c + 2(\nu^*)^2 E^{**} \tag{3.17}$$

$$E^{**} = \left[\frac{(\pi(B^2+L^2)/2BL)-1+\nu}{E} + \eta\frac{(1+\nu_c)(1-2\nu_c)}{E_c} + (1-\eta)\frac{(1+\nu)(1-2\nu)}{E} \right]^{-1}$$

Predictions obtained by the variational method using the linear elastic behavior of the constituents of RS have been recently incorporated in a global methodology of design of column-reinforced foundations detailed in Chapter 4 (Bouassida, 2012).

3.5.3 Some Comments

It is noted that the lower bounds given by Equations (3.16) and (3.17) do not depend on the ratio R/H_c. Moreover, since they depend on the IAR they are applicable for any location of the columns under the rigid raft.

In the case of zero Poisson's ratios ($\nu = \nu_c = 0$), the lower bound in Equation (3.16) reduces to the apparent Young's modulus of RS given by Equation (3.7). This is a well-known estimate, referred to as the *homogenized modulus*, and can also be derived when considering a zero stress field in Zone (I) and one-dimensional stress fields in Zones (II) and (III). In this particular case, the benefit of the confinement provided by the soil in Zone (I) vanishes. Compared to the estimations given by Equations (3.16) and (3.17), a better approximation of the apparent Young's modulus is obtained from the rational variational approach herein.

As an illustration, consider a circular raft with E = 5,000kPa and Q_{all}/A = 100kPa. An estimate of the upper bound of the settlement (Figure 3.6) is a function of Young's modulus of the column material and a prescribed IAR.

The prediction of the settlement reduction can be evaluated by introducing the reinforcement factor Γ defined as the ratio between the apparent Young's

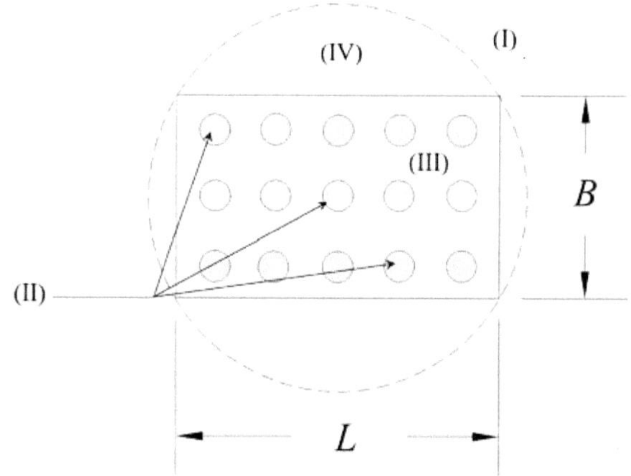

Figure 3.5 Four-zone stress field: variational method, rectangular foundation of length L and width B

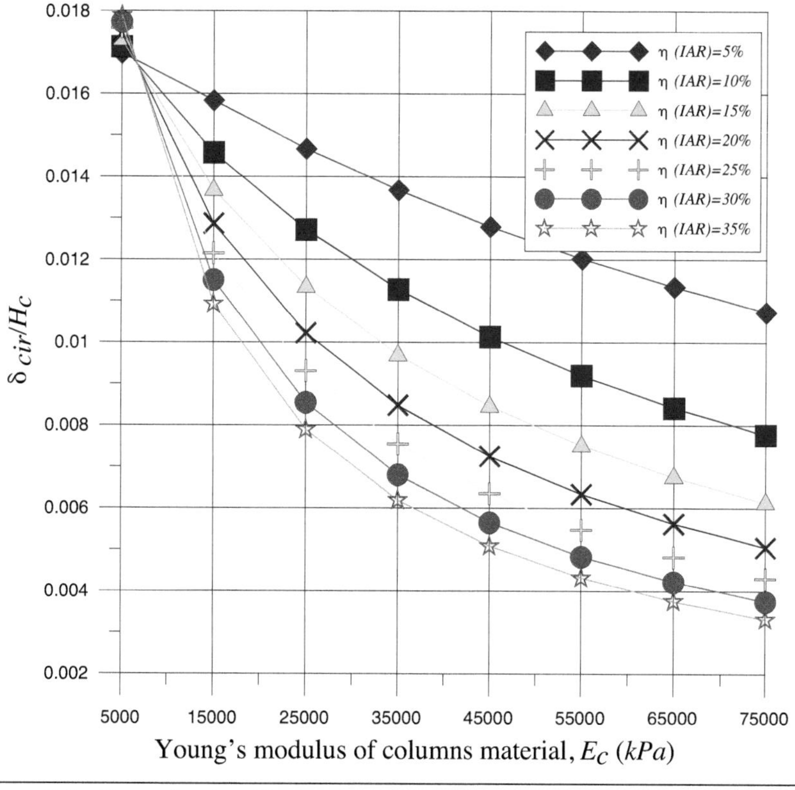

Figure 3.6 Normalized settlement of RS as a function of Young's modulus of column material and fixed IAR (circular rigid raft)

modulus of the RS and that corresponding to the initial soil predicted by the method in question, i.e., Chow's contribution, variational method, Balaam and Booker's method, and the homogenized modulus, i.e.,

$$\Gamma = \frac{E_{app}(\eta \succ 0)}{E_{app}(\eta = 0)} \tag{3.18}$$

Bouassida et al. (2003a) have presented practical design charts that show the variation of reinforcement factor as a function of the normalized Young's modulus of the column material E_c/E. Figure 3.7 presents this variation for a circular raft and soil data: $Q_{all}/A = 100kPa$; $E = 5{,}000kPa$ and $E_c/E = 1 - 15$.

If the upper bound of the settlement of the RS predicted by the variational method is taken as an allowable value, it therefore refers to a maximum IAR (Bouassida 2012). As such, it follows a conservative design from the verification of the allowable settlement of the CRF.

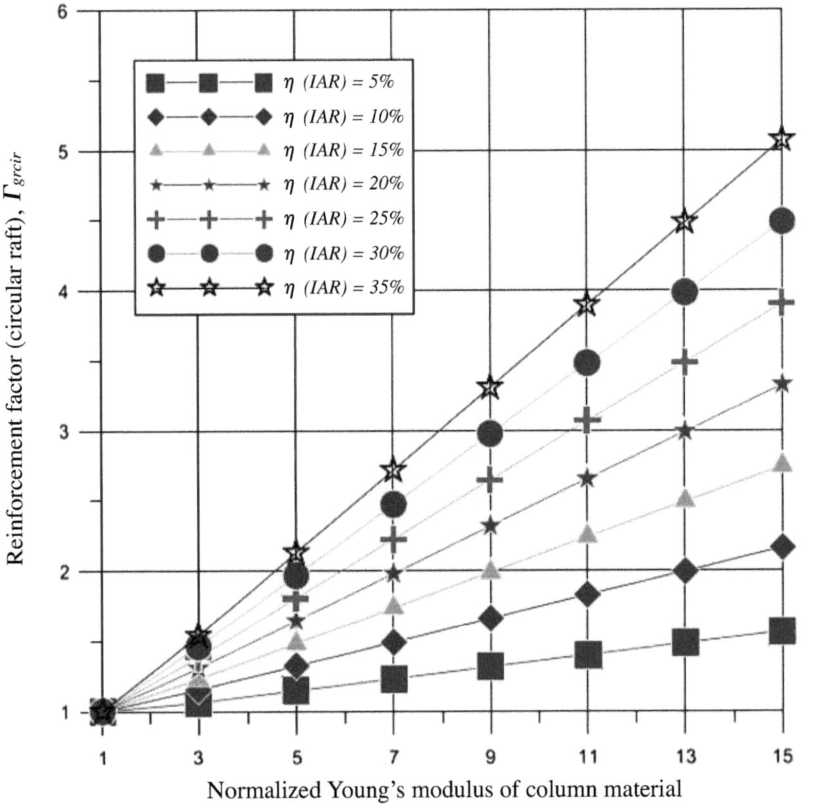

Figure 3.7 Reinforcement factor in function of the normalized Young's modulus of column material for circular rigid raft ($\nu = 0.4$ and $\nu_c = 0.2$)

Bouassida et al. (2003b) have shown that by performing the variational approach in linear elasticity, Chow's (1996) method of analysis leads to the lowest estimation of settlement due to the assumed zero horizontal displacement everywhere in the RS mass.

3.5.4 Homogenization Approach

The homogenization approach is applicable when the reinforcing columns are installed in a regular pattern with an axis-to-axis spacing small enough when compared to the dimensions of the loaded foundation and the thickness of the reinforced layer(s). In the homogenization approach, the initial problem of a heterogeneous-periodical structure is reduced to a homogenized associated problem that is defined for the same structure and subjected to the same loading conditions (de Buhan, 2001). Although the initial soil and column material are assumed isotropic media for the initial problem, it is demonstrated that the homogenized medium has anisotropic characteristics (strength or stiffness properties). This section briefly summarizes the main contributions which have adopted the homogenization approach for the prediction of settlement of a CRF.

Schweiger and Pande (1986) assumed that the stiffness tensor of the homogenized material is a weighted average of the stiffness tensor of column material and the stiffness tensor of the initial soil, as written in Equation (3.7) for the apparent Young's modulus. Anisotropic properties of the equivalent homogeneous material are then derived on the basis of continuity of stress vectors between the initial soil and columns media.

Canetta and Nova (1989) also adopted the same weighted average stiffness tensor considered by Schweiger and Pande (1986), but they derived anisotropic properties by imposing equilibrium equations and compatibility of strains at the interface between the column and the initial soil media. The constitutive law governing the reinforced medium was derived under incremental form but restricted to the plane strain condition, which only applies to the reinforcement by a trench. The authors considered an elastoplastic strain-hardening model for both the initial soil and column material to predict the behavior of a circular raft on soft soil reinforced by vibro-floated sand. Anisotropic properties were considered in this analysis.

Lee and Pande (1998) used the homogenization technique to establish the equivalent material properties for a 3D stone CRF. Critical state Cam Clay and Mohr Coulomb models were considered for the initial soil and column material, respectively, to predict the load-settlement curve of circular footing resting on stone CRFs. This prediction was from an elastoplastic incremental procedure implemented in an axisymmetric FE code, which took account of the initial stresses condition corresponding to columns installed by the replacement method. A scaled test model served to validate the numerical predictions which revealed good agreement with experimental records.

Ben Said et al. (2004) used the homogenization method in a linear elasticity framework, and by employing the variational approach, they were able to determine the homogenized elastic characteristics of the RS as an equivalent anisotropic medium. The predicted settlements from their method were similar to the predictions obtained by the direct approach of the variational method detailed in Section 3.5.

Abdelkrim and de Buhan (2007) have applied an elastoplastic homogenization method to a soil reinforced by regularly distributed columns. An approximate solution to this problem was adopted to investigate the homogenized behavior of the RS subjected to oedometer conditions. A prediction of the settlement of an RS foundation is presented that takes account of the initial stress field generated by gravity. A simplified calculation procedure, in which the soil remains elastic, was proposed, leading to an analytical closed-form expression for the load-settlement curve.

Ghazi et al. (2010) have implemented an elastoplastic homogenization method for periodic media by adopting a simplified constitutive law for a stone column-RS. The closed-form expressions derived for such a constitutive law allow for its implementation into a FE-based numerical procedure. The computational code is then used to simulate the response of RS by a group of floating columns, expressed in terms of load-settlement curves drawn up to the ultimate bearing capacity of the RS.

3.6 ASSESSMENT OF SETTLEMENT PREDICTIONS BY SCALED AND FULL-SCALE MODELS

This section presents comparisons between some recorded field data and predictions made by the variational method detailed in this chapter, as well as by several methods using the linear elastic behavior to predict the settlement of a foundation built on soil reinforced by columns. Two case histories, one, an oil storage tank and the other, an embankment—built on stone CRFs—are analyzed and discussed.

3.6.1 Oil Storage Tank

This case history was reported by Watt et al. (1967) and is of an oil storage tank that was built on a soft clay layer of 6 m thickness, reinforced by end-bearing columns of 1 m in diameter that were installed in a triangular pattern. The axis-to-axis spacing between columns was 1.9 m which corresponds to an IAR of 33 percent. Initial soil characteristics were: $E = 2000kPa$ and $C = 50kPa$. The applied load by the foundation was assumed as uniform: $Q_{all}/A = 150kPa$. Characteristics of the column material were: $E_c = 10E$ and $\varphi_c = 40°$. No data is available for Poisson's ratio for the initial soil or the column material. Accordingly, Fessi-Guetif (2004) performed a parametric study by considering two sets of values: $\nu = \nu_c = 0.33$ and $\nu = 0.45$; $\nu_c = 0.20$ and presented two charts showing the variation of the reinforcement factor Γ, defined in Equation (3.18) as a function of the IAR. Predictions were obtained for IAR = 0.33 by using several methods of design, i.e., Greenwood (1970), Balaam and Booker (1981), Priebe (1995), Chow (1996), and Bouassida et al. (2003a), of the law of mixture and measured data.

As shown in Figures 3.8 and 3.9, it can be noted that the prediction of settlement reduction by Bouassida et al. method's (2003a) is in fair agreement with recorded data, in terms of mean value.

Further, it is confirmed that Greenwood's method leads to a quite conservative prediction of settlement reduction, while Chow's assumption leads to the highest prediction of settlement reduction.

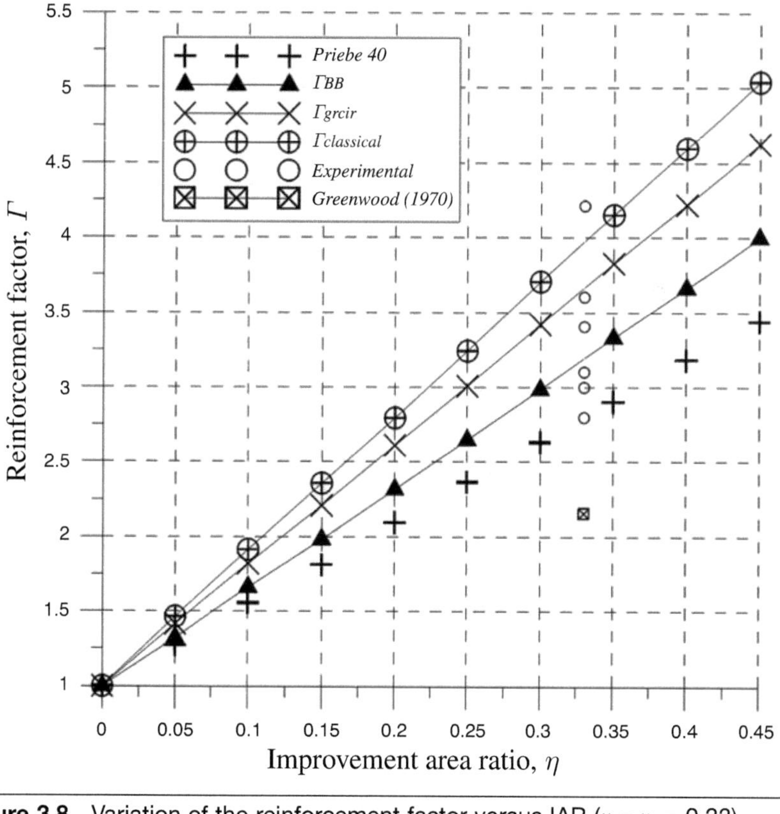

Figure 3.8 Variation of the reinforcement factor versus IAR ($\nu = \nu_c = 0.33$)

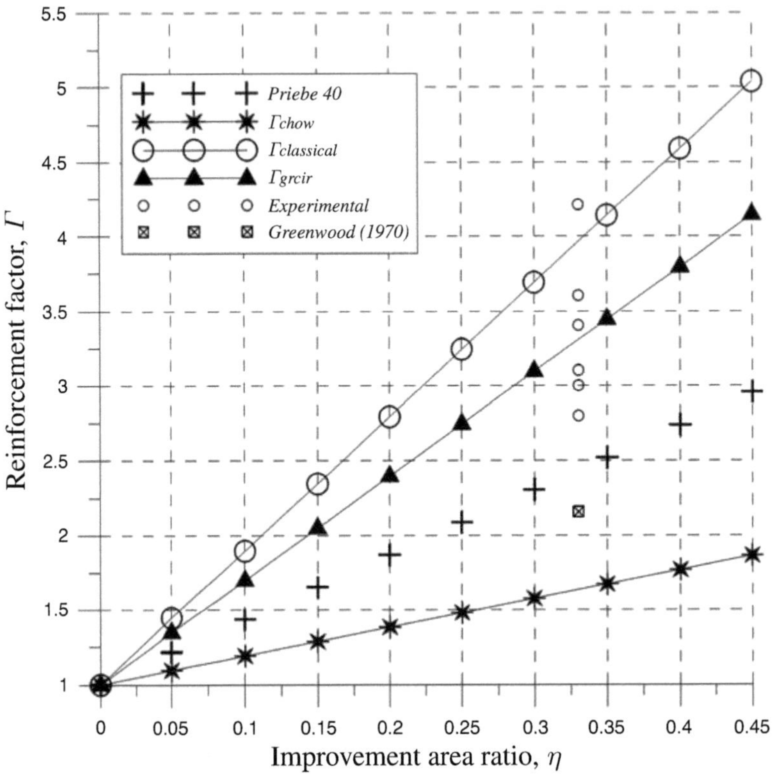

Figure 3.9 Variation of the reinforcement factor versus IAR ($\nu = 0.4$ and $\nu_c = 0.2$)

In addition, one can see that the prediction of settlement reduction is underestimated by the method of Balaam and Booker (1981); and the reinforcement factor as predicted by Priebe's method is overestimated.

Further, Figure 3.9 shows that Poisson's ratios of the initial soil and column material also have significant influence on the settlement prediction.

3.6.2 Storage Facility of Protected Bullets in Mounded Banks in Tunisia

British Gas-Tunisia built, in the industrial area of Gabes Ghannouch (southeast of Tunisia), a storage facility composed of two bullets of butane and five bullets of propane protected in mounded banks. Figure 3.10 schematizes the cross section of the whole integrated embankment resting on multilayers of soil profile that is reinforced by floating stone columns. Geotechnical properties of soil layers were adopted on the basis of measured cone penetration test (CPT) values

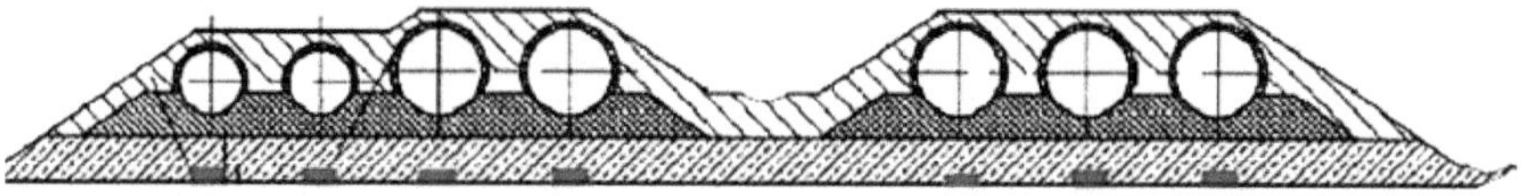

Figure 3.10 Cross section of embankment loading of Ghannouche case study (with permission from Editor E. Hamdi)

during soil investigation and laboratory test results conducted for the project. Stone column reinforcement was essentially decided to reduce unallowable settlement as predicted under an applied embankment load of 120 kPa. More precisely, required stability was agreed on the basis of an allowable settlement of 4 cm over 15 years in post-construction of the storage facility (El Ghabi et al. 2010). Hence, significant reduction of settlement associated to the prescribed margin of security has led to the installation of floating stone columns of 11 m in length, embedded in a fine-to-medium sand layer. Stone columns with a 0.9 m diameter were installed in an equilateral triangular pattern of 1.9 by 2.2 m with IAR = 16%. Adopted characteristics of column material were: $\gamma_c = 20kN/m^3$ $E_c = 60{,}000kPa$, $\nu_c = 0.33$ and $\varphi_c = 40°$.

Table 3.4 summarizes the predictions of linear elastic settlement obtained by *Columns 1.01* software for the unreinforced soil and the RS (Bouassida and Hazzar, 2012). Table 3.4 shows that the French method (2005) and Balaam and Booker (1981) methods give a quite similar prediction of total settlement. Although both of these methods do not take into account the improvement of the initial soil, the settlement reduction that resulted from the installation of floating stone columns is nearly one third. More importantly, the use of *Columns 1.01* software has confirmed that the optimized IAR, which verifies the allowable settlement of

Table 3.4 Predictions of linear elastic settlement of embankment on RS by floating stone columns

Layers	Thickness (m)	Settlement of unreinforced soil (cm)	Settlement of reinforced soil (cm)	
			Balaam and Booker (1981)	**French method (2005)**
Gypseous sand	2.5	0.5	0.26	0.25
Silty sand	6.5	6.83	1.34	1.53
Medium cemented sand (1)	5.5	0.54	0.37	0.41
Medium cemented sand (2)	7.5	3.58	1.18	1.35
Compacted fine sand	6.0	0.41	0.41	0.41

4 cm, is 15.39%. This prediction is quite close to the practiced reinforcement for the project that was 16%.

3.7 CONCLUSIONS

In this chapter the methods for settlement prediction have been reviewed where the linear elastic behavior is adopted for both the initial soil and the columns. As a practical result, the settlement reduction, being the principal aim of reinforcement by columns, is expressed in terms of the IAR.

The settlement prediction of a rigid foundation resting upon an RS by end-bearing columns has been investigated by the use of a variational approach in linear elasticity, which provided a lower bound of the apparent Young's modulus of RS from which an upper bound settlement estimation is deduced. It follows a secured estimation of settlement of loaded rectangular and circular rigid foundation resting on RS by a group of columns in arbitrary arrangement. The results have been presented in the form of convenient design charts; and these can be used to optimize the IAR for a prescribed load and a given allowable settlement of the foundation.

The test results from two cases histories were used to compare several methods of settlement prediction. It has been verified that the settlement prediction by Priebe's method is more secured in comparison to predictions made by several methods adopting the linear elastic behavior for constituents of the RS. However, the predictions by those linear elastic methods were in good agreement with the measured settlement for different case histories.

Finally, it should be noted that little interest has been accorded for the prediction of settlement of foundations resting on compressible clays reinforced by floating columns. Such a case of study deserves special attention for determining the evolution of settlement of unreinforced compressible layers since it controls the long-term behavior of the foundation. Chapter 4 introduces an optimized design of CRF and gives insight into the estimation of settlement of compressible clays reinforced by floating columns.

Appendix 3.1 Variational Approach in Linear Elasticity

Construction of a Three-Zone s.a. Stress Field

Figure 3.5 shows the construction of this type of stress field in three complementary areas within the RS mass. Components of the stress tensor are expressed in the local basis $(\underline{e}_r;\underline{e}_\theta;\underline{e}_z)$

Zone (I): the initial soil outside the cylinder with radius R and height H_c:

$$\underline{\underline{\sigma}}' = w_1 \left(\frac{R}{r} \right)^2 \left(\underline{e}_\theta \otimes \underline{e}_\theta - \underline{e}_r \otimes \underline{e}_r \right) \tag{3.19}$$

Zone (II): the set of all columns with a total volume: $\eta \pi R^2 H_c$:

$$\underline{\underline{\sigma}}' = -w_1 \left(\underline{e}_\theta \otimes \underline{e}_\theta + \underline{e}_r \otimes \underline{e}_r \right) - w_2 \underline{e}_z \otimes \underline{e}_z \tag{3.20}$$

Area (III): the initial soil between the columns located under the foundation of volume: $(1-\eta)\pi R^2 H_c$:

$$\underline{\underline{\sigma}}' = -w_1 \left(\underline{e}_\theta \otimes \underline{e}_\theta + \underline{e}_r \otimes \underline{e}_r \right) - w_3 \underline{e}_z \otimes \underline{e}_z \tag{3.21}$$

The family of stress fields expressed by Equations (3.19), (3.20), and (3.21) where w_1, w_2, and w_3 are real parameters is s.a. Substituting Equations (3.19), (3.20), and (3.21) in (3.10), (3.11), (3.12), (3.13), and (3.14), the function of complementary energy is calculated.

Then, applying the principle of minimum complementary energy via Inequality (3.15), one obtains, after minimization over parameters w_1, w_2, and w_3 the lower bound expression of the apparent Young's modulus of the RS (Bouassida et al., 2003a).

REFERENCES

Abdelkrim, M. and de Buhan, P., An elastoplastic homogenization procedure for predicting the settlement of a foundation on a soil reinforced by columns. *Europ. J. of Mech. and Solids*, 26, 2007, 736–757.

Ambily, P. and Gandhi, S. R., Behavior of Stone Columns Based on Experimental and FEM Analysis. *J. Geotech. Geoenviron. Eng.*, ASCE 133 (4), 2007, 405–415.

Balaam, N. P. and Booker, J. R., Analysis of rigid rafts supported by granular piles. *Int. J. of Num. & Analyt. Meth. in Geomech.*, 5 (4), 1981, 379–403.

Balaam, N. P. and Poulos, H. G., The behavior of foundations supported by clay stabilized stone columns. Proc. 8th Europ. Conf. on SMFE, Helsinki, May, 1983, 1, 199–204.

Balaam, N. P. and Booker, J. R., Effect of stone column yield on settlement of rigid foundations in stabilized clay. *Int. J. of Num. & Analyt. Meth. in Geomech.*, 9, 1985, 331–351.

Balaam, N. P. and Booker, J. R., Program Colany. Stone columns settlement analysis. User Manual; The University of Sydney, 1987 (Australia).

Ben Said, B., Bouassida, M. and de Buhan, P., Estimation par une méthode d'homogénéisation du tassement d'une fondation sur un sol renforcé par colonnes. *Proc. Int. Symp. on Ground Improvement, ASEP-GI'04*, Lab. Central Pts. et Ch., Paris, 2004, 59–66.

Bergado, D. T., Anderson, L. R., Miura, N. and Balasubramaniuam, A. S., Soft Ground Improvement in Lowland and Other Environments. ASCE Press. New York, 1996.

Bouassida, M., Etude expérimentale du renforcement de la vase de Tunis par colonnes de sable—Application pour la validation de la résistance en compression théorique d'une cellule composite confinée. *Revue Française de Géotechnique*, 75 (2), 1996, 3–12.

Bouassida, M., Guetif, Z., de Buhan P. and Dormieux, L., Estimation par une approche variationnelle du tassement d'un sol renforcé par colonnes. *Revue Française de Géotechnique*. 102 (1), 2003a, 21–29.

Bouassida, M., Guetif, Z., de Buhan, P. and Dormieux, L., Settlement estimation of a foundation resting on a soil reinforced by columns. *Proc. 13th African Reg. Conf. on Soil Mech. & Geotech. Eng.*, 2003b, Marrakech (Morocco) 467–473.

Bouassida M., Design of columnar-reinforced foundations. Invited paper, Int. Conf. on Ground Improvement and Ground Control: ICGI 2012, 30 Oct–2 Nov, 2012. University of Wollongong (Australia), Indraratna et al. (Eds), 1, 599–608.

Bouassida, M. and Hazzar, L., Novel tool for optimized design of reinforced soils by columns. Ground Improvement: Proc. ICE, London 165 (1), 2012, 31–40.

Broms, B. B., Lime columns in theory and practice. Proc. Inter. Conf. Soil Mech. Mexico, 1982, 149–165.

Broms, B. B., Lime and lime-cement columns: Summary and Visions. Proc. 4th Int. Conf. on Ground Improvement Geosystems, Keynote Lecture, Helsinki, 7th–9th June, 2000, 43–93.

Chow, Y. K., Settlement analysis of sand compaction piles. *Soils and Foundations J.*, 36 (1), 1996, 111–113.

Canetta, G. and Nova, R., A numerical method for the analysis of ground improvement by columnar inclusions. *Computers & Geotechnics*, 7, 1989, 99–114.

Datye, K. R., Settlement and bearing capacity of foundation system with stone columns, C. R. Proc. Int. Symp. Soil and rock improvement techniques including geotextiles, reinforced earth and modern piling methods, AIT, Bangkok, 1982, 1.1-A. 1–27.

de Buhan, P., Homogénéisation des milieux périodiques élastiques, Cours de DEA de l'École Nationale des Ponts et Chaussées, 2001, Paris.

Dhouib, A., Au sujet de l'article: Recommandations sur la conception, le calcul, l'exécution et le contrôle des colonnes ballastées sous bâtiments et ouvrages sensibles au tassement. *Revue Française de Géotechnique*, 117 (4), 2006, 63–66.

Dhouib, A. and Blondeau, F., Colonnes ballastées. Edit. Presses des Ponts et Chaussées, Paris, 2005. 142–144.

El Ghabi, B., Hermges, T., Lambert, S., Ellouze, S., Bouassida, M. and Mejri, A., Comparison between predicted settlement of a group of stone columns foundation and in situ records. Proc. 2nd Int. Conf. on Geotech. Eng., Bouassida et al. (Eds), October 25–27, 2010, Hammamet (Tunisia), 255–264.

Ellouze, S., Bouassida, M., Hazzar, L. and Mroueh, H., On settlement of stone column foundation by Priebe's method. *Proc. of the Institution of Civil Engineers Ground Improvement J.* 163 (2), 2010, 101–107.

Fessi-Guetif, Z., On the prediction of settlement of foundations on reinforced soil by columns. PhD's dissertation (in French), defended on 05 March, 2004. National Eng. School of Tunis, Tunisia.

French Method. Recommandations sur la conception, le calcul, l'exécution et le contrôle des colonnes ballastées sous bâtiments et ouvrages sensibles au tassement. *Revue Française de Géotechnique*, 111 (2), 2005, 3–16.

Ghazi H., de Buhan, P. and Abdelkrim, M., Finite element implementation of a homogenized constitutive law for stone column-reinforced foundation soils, with application to the design of structures. *Computers & Geotechnics*, 2010, 40–49, Elsevier.

Ghionna, V. and Jamiolkowski, M., Colonne di ghiaia, X ciclo di conferenze dedicate ai problemi dei terreni e ingegneria delle fondazioni metodi di miglioremento dei terreni, politechnico di Torino Ingegneria, atti dell'istituto di scienza delle costruzioni, 1981, 507.

Goughnour, R. R., Settlement of vertically loaded stone columns in soft ground. Proc. Special session, 8th Europ. Conf. on Soil Mech. and Found. Eng., 1983, 2, Helsinki,

Goughnour, R. R., and Bayuk, A. A., Analysis of stone column soil matrix interaction under vertical load. Proc. Int. Conf. on Soil Reinforcement, l'École Nationale des Ponts et Chaussées, Paris, 1979, 271–277.

Greenwood, D. A. Mechanical improvement of soils below ground surface. Symp. on Ground Engineering, Institution of Civil Engineers, London, U.K., 1970, paper II: 09–20.

Hughes, J. M. O., Withers, N. J. and Greenwood, D. A., A field trial of reinforcing effects of stone columns in soil. *Géotechnique*, 25 (1), 1975, 31–44.

Lee, J. S. and Pande, G. N., Analysis of behavior of stone columns and lime columns, *Computers and Geotechnics*, 20 (1), 1998, 47–70.

Mitchell, J. K. and Huber, T. R., Performance of a stone-column foundation. *J. Geotech. Engin., ASCE*, 1985, 111, 205–223.

Poorooshasb, H. B. and Meyerhof, G. G., Analysis of behavior of stone columns and lime columns. *Computers and Geotechnics*, 1997, 1 (20), 47–70.

Priebe, H. J., Estimating settlements in a gravel column consolidated soil. *Die Bautechnik*, 53 (5), 1976, 160–162 (in German).

Priebe, H. J., The design of vibro replacement. *Ground Engineering* 28 (10), 1995, 31–37.

Priebe, H., Program manual GRETA ground improvement. 2003, GEOStat.

Rao, G. and Ranjan, G., Settlement analysis of skirted granular piles. *J. of Geotech. Eng. ASCE*, 111 (6), 1264–1281.

Salençon, J., Mécanique des milieux continus, tome I: Elasticité-Milieux curvilignes. Editions, Ellipses, 1988, Paris.

Schweiger, H. F. and Pande, G. N., Numerical analysis of stone columns supported foundations. *Computers and Geotechnics*, 1986, 2 (6), 347–372.

Six, V., Mroueh, H., Shahrour, I. and Bouassida, M., Numerical analysis of elastoplastic behavior of stone column foundation. *Geotechnical Geological Eng*, 2012, Springer. 30:813–825.

Soyez, B., Méthodes de dimensionnement des colonnes ballastées. *Bulletin de Liaison des Laboratoire des Ponts et Chaussées*, 135, Janvier-Février 1985, Paris, 35–51.

Van Impe, W. and De Beer, E., Improvement of settlement behavior of soft layers by means of stone columns. Proc. 8th Europ. Conf. Soil Mech. and Found. Eng. Helsinki, 1983, May, 309–312.

Watt, A. J., de Boer, J. J. and Greenwood, D. A., Loading tests on structures founded on soft cohesive soils strengthened by compacted granular columns. Proc. 3rd Asian Conf. of Soil Mech. and Found. Eng., 1, Haifa, 1967, 248–251.

4

NOVEL METHODOLOGY FOR DESIGN OF COLUMN-REINFORCED FOUNDATIONS

ABSTRACT

The design of foundations on soft ground reinforced by columns usually involves two important verifications, specifically, checking for adequate bearing capacity and acceptable settlement performance. This chapter details a comprehensive methodology for determining the optimized portion of the ground area that should be improved by the installation of columns. The optimization is required in order to avoid an overly conservative design and, consequently, the use of uneconomical quantities of material to construct the column reinforcement. The basis of the suggested methodology consists of first estimating the minimum improvement area ratio (IAR) required to ensure attainment of the required design bearing capacity of the reinforced soil (RS), and then determining an upper bound or maximum value of IAR by considering the issue of allowable settlement. Optimization is then performed on IAR within the range defined by these bearing capacity and settlement limits. Analysis of three case studies provides an illustration of the implementation of this novel design methodology, which has been incorporated in software recently developed to assist in the design of soil foundations reinforced by columns and to provide cost-effective solutions for this type of foundation.

4.1 INTRODUCTION

It has been well explained in the previous chapters that the reinforcement of weak soils by stiffer columns has numerous benefits, including the potential to increase the bearing capacity, reduce settlements under working loads, accelerate consolidation of the soft soil by acting as a drainage pathway (whenever free-draining material is used to construct the columns), and reducing the risk of liquefaction, especially in the case of saturated loose sands. The cost of most schemes involving column-reinforced foundations (CRF) that are constructed using stone columns, sand compaction piles, or deep soil mixing techniques is essentially controlled by the volumetric fraction of material incorporated into the host soil (or native ground). This fraction is referred to as the IAR, which is defined by the total cross-sectional area of the columns divided by the gross area of the loaded foundation (Chapter 2). Depending on the adopted technique of column reinforcement, the value of the IAR generally falls within specific ranges, as detailed in Chapter 1.

The design of foundations on ground reinforced by columns usually requires verifications: first, of adequate bearing capacity, and second, acceptable settlement performance. The design can also involve consideration of the acceleration of any consolidation settlement of the host soil when the columns behave like vertical drains, as well as the liquefaction potential of the host soil which is relevant mostly for loose saturated sands.

Prior methods suggested for the design of the CRF are mainly based on the unit cell model (UCM), which provides the particular advantage of allowing validation of analytical and numerical predictions with experimental records. The latter have usually been obtained from triaxial tests (e.g., Barksdale and Bachus, 1983; Ghionna and Jamiolkowski, 1981; Bouassida, 1996; Poorooshab and Meyerhof, 1997; Ambily and Ghandi, 2007). It is assumed that lateral deformation at the boundary of the UCM is zero in conformity with oedometer conditions. The main advantage of this laboratory model is the assessment of theoretical results in regard to bearing capacity, settlement, and acceleration of consolidation of a compressible soil reinforced by drained column material. Accordingly, Bouassida (1996), Ambily and Ghandi (2007), and others conducted experimental investigations adopting the UCM in order to study the behavior of soil reinforced by columns.

The isolated column model (ICM) has also been used to predict the ultimate bearing capacity and settlement of the CRF. These predictions have been compared with data recorded from full-scale loading tests, usually performed in field trials conducted prior to installation of the majority of columns (Bergado and Lam, 1987; El Ghabi et al., 2010). The ICM is therefore a useful tool for the validation of in situ conditions. However, it does assume that a single column is surrounded by the host soil and, therefore, restrained in some way by that host soil. But in particular, because it is also assumed that the surface loading is applied directly to the column only, it can effectively be considered as a special case of full soil reinforcement, i.e., a full replacement configuration (IAR = 100%). This severe limitation of the ICM means that the influence of the IAR on the performance of the CRF cannot be studied using this approach. It also means that the use of the ICM, as adopted in the French Method (2005), generally leads to an overestimated prediction of the ultimate bearing capacity of stone column foundations compared to predictions given by other methods (Ellouze et al., 2010). Accordingly, adopting three-dimensional modeling of CRF would seem more appropriate, in order to obtain representative predictions of the role of the IAR as a key parameter in design.

Further, the existing methods of design of the CRF are often aimed at obtaining a unique verification of either the bearing capacity (less likely) or the settlement (more likely) as the controlling factor in design. As such, these methods of design have been formulated for a unique method of column installation, e.g., stone columns (Priebe, 1995; French Method, 2005) or the deep soil mixing method (Kitazume et al., 1996; Broms, 2000). In these contributions, optimization of the quantity of required column material was not a matter of first importance. Therefore, the values of the IAR were usually dictated by experience that was gained from existing CRF projects with a focus on the reduction of settlement and perhaps acceleration of that settlement, which are factors mainly related to stone column reinforcement.

In this chapter, a comprehensive methodology for the design of the CRF is suggested that considers satisfying the requirements of both bearing capacity and settlement. The suggested methodology is based on research results that have been obtained in rigorous frameworks involving 3D modeling of the CRF as end-bearing and floating columns. In this approach, the constituents of the CRF (i.e., the weak native soil and the reinforcing columns) are assumed to be different homogeneous and isotropic materials. First, limit analysis is used to determine the ultimate bearing capacity of the RS by considering the strength criteria of the initial soil and the column material, with both obeying the rigid perfect plastic Mohr Coulomb model (Bouassida et al., 1995). Second, the variational method in linear elasticity is used to predict the settlement of the RS (Bouassida et al., 2003).

The suggested methodology requires three steps to derive an optimized value of IAR for any given situation. The IAR was found to be the key parameter for the design because the reinforced ground is considered by a group of columns. Once the optimized IAR is determined for a given column technique, the diameter of the inclusions is provided within a certain range (as an example, for stone columns it is between 0.8 m and 1.2 m), the spacing between columns is then deduced for a chosen grid pattern. This design procedure is already programmed in *Columns 1.01* software (Bouassida and Hazzar, 2012).

The reductions in project costs achieved by adopting the suggested methodology of design is evidenced by three case histories: a tank foundation resting on soft clay reinforced by end-bearing stone columns; an embankment founded of soft clay reinforced by either floating lime-cement treated columns or by full substitution of the host soil either by compacted sand or cement stabilized host soil; and an embankment founded on soft clay reinforced by floating columns installed by the deep mixing method (DMM).

4.2 PROBLEM STATEMENT

The first stage in the design process of the CRF involves verification of bearing capacity, which constitutes a necessary condition for the stability of the CRF. Thereafter follows verification of the settlement criteria for the CRF, which is also a necessary condition that must be satisfied in order to achieve an adequate design. Reinforcement of the host soil by columns is achieved by vertical inclusions—either end-bearing or floating—of length H_c, installed under the loaded area A, referred to as the *Foundation* in Figure 3.1. The locations of the columns having circular cross sections can be arbitrary and so may their diameters (Figure 1.7). The total cross-sectional area of columns is denoted A_c. The IAR, denoted by the symbol η, is then given simply as:

$$\eta = \frac{A_c}{A} \tag{4.1}$$

Verification of bearing capacity involves estimation of the unit weights and shear strength characteristics of the component soils (i.e., friction angle and cohesion) of the initial host soil and the column material. The settlement is estimated by using Young's moduli and Poisson's ratios of the column material (E_c, ν_c) and the weaker host soil (E, ν). Because introduction of the column material is intended to reinforce the host soil, and thus increases its overall effective stiffness, it generally follows that $E_c > E$.

The first step of the methodology is to consider lower and upper bounds of the ultimate bearing capacity of the CRF, detailed in Chapter 2, in order to determine the minimum value of IAR (η_{min}) that complies with the maximum allowable bearing capacity of the CRF (the design capacity).

In the second step, the settlement of the CRF is estimated within the framework of linear elasticity, within which application of variational principles (detailed in Chapter 3) enables an assessment of the lower bound of the apparent Young's modulus that is associated with the maximum allowable settlement.

Given a specified maximum allowable settlement of the CRF, an upper bound or maximum value of IAR (η_{max}) is then identified.

Finally, the value of IAR (η) is optimized within the range defined by these first two steps ($\eta_{min} \le \eta \le \eta_{max}$) by considering the specified settlement criteria and various models for predicting the settlement of the CRF. In this way an optimum value of IAR (η_{opt}) is determined that does not overestimate the quantity of reinforcing column material.

The methodology to be detailed in the following sections applies for different types of structures (e.g., rigid rafts or footings and embankments) for which the settlement at the surface of the RS, δ_r, is assumed to be uniform. The proposed design methodology is then based on the verifications of bearing capacity and settlement which are conducted for the short-term condition that is the first requirement for the stability of the CRF. Once the optimized area ratio is identified, it should be checked as to whether or not it fits within the prescribed ranges given in Chapter 1. If so, a finite element procedure can be used to predict the long-term behavior of the structure founded on the reinforced column foundation.

4.3 BEARING CAPACITY OF CRF

The limit analysis framework is usually considered by performing a static analysis to derive a lower bound of the ultimate bearing capacity of the CRF and a kinematic analysis to determine its upper bound. Based on existing results which have been obtained by Bouassida et al. (1995) and Bouassida and Porbaha (2004), predictions of the minimum values of the IAR will be detailed shortly. Considering that the weak soil is almost purely cohesive, the behavior of the column material is either frictional (such as that of stone, gravel, or sand, for which cohesion can usually be neglected) or purely cohesive (such as soft clay treated by cement, lime, or a combination of the two). Hence, for calculation purposes, two main categories of column material, i.e., cohesive-frictional and purely cohesive, are used to study the most-used techniques, which are stone columns and deep mixing.

4.3.1 Case of Stone Columns

By adopting the static approach of limit analysis, a lower bound of the ultimate bearing capacity of the CRF, Q^-_{ult}, is determined for a purely cohesive soil reinforced by a group of end-bearing columns made up of cohesive-frictional material. This lower bound solution obtained by Bouassida et al. (1995) may be written as:

$$Q^-_{ult} / A = \sigma_{ult,rs} = (1-\eta)\sigma_{ult,s} + \eta\sigma_{ult,c} \qquad (4.2)$$

In Equation (4.2) the vertical stresses induced within the column material and the host soil, respectively, i.e., $\sigma_{ult,c}$ and $\sigma_{ult,s}$, may be expressed in terms of the friction angle φ_c and cohesion C_c of the column material and the cohesion (undrained shear strength) of the host soil C. Explicit expressions for the ultimate load Q^{-}_{ult}, are provided in Appendix 4.1 for the case of a purely cohesive soil reinforced by cohesive-frictional column material. Explicit expressions for the individual terms $\sigma_{ult,c}$ and $\sigma_{ult,s}$, can easily be derived from this detail.

An appropriate value of the factor of safety against yielding, denoted F_{rs}, is adopted and regardless of the failure mode of the RS, the allowable bearing capacity of the CRF, denoted as $\sigma_{all,rs}$, is defined by:

$$\sigma_{all,rs} = \frac{(1-\eta)\sigma_{ult,s} + \eta\sigma_{ult,c}}{F_{rs}} \tag{4.3}$$

As a first approximation, the predicted lower bound of the bearing capacity can usually be associated with a factor of safety within the range $1 < F_{rs} < 2$. The allowable bearing capacity of the CRF, compared to the mean vertical stress induced by the load applied to the foundation, denoted Q_{actual}, should comply with the necessary condition:

$$Q_{actual} / A \leq \sigma_{all,rs} \tag{4.4}$$

Combining Equations (4.3) and (4.4), the minimum value of IAR, denoted as η_{min}, is then identified so that:

$$\eta \geq \frac{F_{rs}(Q_{actual} / A) - \sigma_{utl,s}}{\sigma_{utl,c} - \sigma_{utl,s}} = \eta_{min} \tag{4.5}$$

As can be observed from Equation (4.5), the minimum value of IAR, η_{min}, corresponds to the minimum quantity of column material to be incorporated in the host soil such that the ultimate bearing capacity increases from $\sigma_{ult,s}$ to $\sigma_{ult,rs}$. Otherwise, if $\eta_{min} \leq 0$, the reinforcement is unnecessary since the allowable bearing capacity of the unreinforced soil is sufficient to withstand the applied load exerted by the foundation.

4.3.2 Case of Deep Mixing Columns

In practice, the CRF constructed by the DMM can be modeled reasonably as a relatively uniform clay medium with an undrained shear strength (cohesion) C, reinforced by another purely cohesive (clay-like) column material with undrained shear strength (cohesion) $C_c = K_c C$ (Kitazume et al., 1996). This case of reinforcement by a group of end-bearing columns has been investigated by performing limit analysis using the kinematic approach, from which an upper bound solution of the ultimate bearing capacity was proposed. The assessment of this upper bound solution was found to be in good agreement with experimental data recorded from scaled models. Considering the same unit weight for the soft clay and the cemented column material, it can be shown that the upper bound of the ultimate bearing capacity of the CRF may be expressed as (Bouassida and Porbaha, 2004):

$$Q^{+}_{ult} / CA = 2\left(\sqrt{2} + \sqrt{\left[1+\eta(K_c - 1)\right]\left[2+\eta(K_c - 1)\right]}\right) \tag{4.6}$$

Where K_c = the ratio of the undrained shear strength of the column material to that of the host soft clay. Current experience from the practice of deep mixing indicates that K_c varies in a large range from 10 to 130, while the value of the IAR usually ranges from 0.1 to 0.7 (Broms, 2000). Within these margins for K_c and η, it can be shown that the upper bound given by Equation (4.6) can be approximated by the following linear relationship (Bouassida and Porbaha, 2004):

$$Q^{+}_{ult} / CA = 4\sqrt{2} + 2\eta(K_c - 1) \tag{4.7}$$

It can be verified that the relative error between the functions given by Equations (4.6) and (4.7) is less than 1 percent, which confirms the usefulness of Equation (4.7), from which the minimum IAR was derived. Therefore, Equation (4.7) can be written in the form:

$$\sigma^{+}_{ult,rs} = \sigma^{+}_{ult,s} + \eta\sigma^{+}_{ult,c} \tag{4.8}$$

In Equation (4.8), $\sigma^{+}_{ult,c}$ and $\sigma^{+}_{ult,s}$ denote the ultimate vertical stress components which are derived from the upper bound solution as given by Equation (4.7).

Following the same steps as those followed from Equation (4.3) through Equation (4.5) for the lower bound solution, and considering the upper bound given by Equation (4.8), the minimum acceptable value of the IAR may be written as:

$$\eta \geq \frac{F_{RS}(Q_{actual} / A) - \sigma^{+}_{ult,s}}{\sigma^{+}_{ult,s}} = \eta^{+}_{min} \tag{4.9}$$

Determination of the bearing capacity of a CRF that was constructed using floating columns has been studied by Bouassida et al. (2009). The results obtained indicate that the lower bound solutions for the ultimate bearing capacity of a CRF constructed with end-bearing columns also apply in the case of reinforcement provided by floating columns; the floating columns are at least of a certain length. As demonstrated by Bouassida et al. (2009) this minimum length is usually achieved in most, if not all, practical cases. Therefore, the prediction of the minimum IAR provided previously also remains valid for the cases where the CRF is constructed using floating columns (Bouassida et al., 2009).

4.4 SETTLEMENT OF COLUMN-REINFORCED FOUNDATIONS

4.4.1 Settlement of End-Bearing Columns

From Figure 3.1, when $H = H_c$, the case corresponds to reinforcement by end-bearing columns, for which the total settlement equals that of the RS mass: $\delta = \delta_r$.

A lower bound of the apparent soil Young's modulus of the CRF, denoted as E_{app}, can be determined using the variational methods in linear elasticity. As detailed in Chapter 3, Bouassida et al. (2003) used the principle of minimum complementary energy, which simply states that for all elastic stress states satisfying the boundary conditions, those that also satisfy the equilibrium equations make the complementary energy a local minimum. Applying this principle to a rigid foundation of area A, subjected to an allowable working load, Q_{actual}, which induces an assumed uniform settlement of the CRF, denoted by δ_r, Bouassida et al. (2003), demonstrated that:

$$E_{app} \geq \frac{Q_{actual} / A}{\delta_r / H_c} \tag{4.10}$$

Furthermore, Bouassida et al. (2003) demonstrated that the lower bound of the apparent Young's modulus of the CRF is greater than the so-called *homogenized* Young's modulus given by: $E_{hom} = (1 - \eta)E + \eta E_c$. Accordingly, from Inequality (4.10) a more conservative prediction of the settlement of the CRF, δ_r, is obtained (Bouassida, 2013):

$$\delta_r \leq \frac{(Q_{actual} / A)H_c}{(1-\eta)E + \eta E_c} = \delta_r^+ \tag{4.11}$$

The predicted upper bound estimate of settlement, δ_r^+, obtained from Equation (4.11) is associated with a maximum value of IAR, denoted $\eta_{\max}$. It is then required that the allowable settlement of the CRF, denoted as $\overline{\delta_r}$, should comply with δ_r^+, so that:

$$\overline{\delta_r} \leq \delta_r^+ \tag{4.12}$$

At this stage of the design procedure, as a first check on the settlement of the CRF, Equation (4.11) is used to verify whether the minimum value of IAR, calculated either from Equation (4.5) or from Equation (4.9), complies with the prescribed allowable settlement of the CRF. If so, then the predicted minimum value of the IAR, η_{min}, can be accepted for the final design because both bearing capacity and settlement requirements have been verified. It is noted that this situation is most likely to be encountered in cases involving the improvement of loose sands using the vibro-compaction method. The bearing capacity of loose sands having a friction angle close to 29 degrees is adequate, even when the soil is not reinforced. Therefore, the optimized IAR is only relevant to the verification of allowable settlement. Accordingly, the optimized IAR reduces to that of the maximum IAR.

In cases where the minimum value of the IAR does not satisfy the allowable settlement criterion, the value of the IAR must be increased such that $\eta > \eta_{min}$. This situation usually arises in cases where soft clay is reinforced by stone columns. For different case histories, Bouassida and Hazzar (2012) showed that the minimum IAR does not satisfy the required settlement criterion of soft clays reinforced by stone columns. It was found that the optimized IAR is often higher than the minimum IAR. In fact, due to the high compressibility of soft clays, even when reinforced by end-bearing columns, the allowable settlement is always satisfied only with a greater IAR than the minimum IAR.

In such cases the design then proceeds by combining Inequalities (4.11) and (4.12), to obtain:

$$\eta \leq \frac{(Q_{actual} / A)(H_c / \bar{\delta}_r) - E}{E_c - E} = \eta_{max} \tag{4.13}$$

From Inequality (4.13) a maximum value of IAR, denoted η_{max}, is then identified. This value indicates the limit beyond which the volume of reinforcement material will definitely be overestimated, since a truly conservative value of Young's modulus of the CRF has been considered for the settlement estimation as indicated in Equation (4.11).

Equation (4.13) shows that the maximum IAR essentially depends on the allowable settlement of the reinforced layers and obviously on the length of the columns. Therefore, for end-bearing or floating columns, the maximum IAR remains unchanged. This parameter does not depend on the thickness of the unreinforced layer, for which the long-term settlement, especially in the case of highly compressible layers, should also be allowable.

Inequalities (4.5) or (4.9) and (4.13) provide bounds on the optimum value of IAR, i.e.:

$$\eta_{min} \leq \eta \leq \eta_{max} \tag{4.14}$$

Inequality (4.14) provides the range of values of IAR that satisfy both the allowable bearing capacity and settlement criteria for the CRF. Within this range, the optimized value of IAR, η_{opt}, may then be determined as a function of the allowable settlement adopted for the design.

An iterative procedure is then carried out within the range: $[\eta_{min} : \eta_{max}]$. This is achieved by incrementing η in suitably small steps within the defined range, commencing at the minimum value η_{min}, and then predicting the settlement corresponding to each particular value of η. The suggested algorithm used to perform this iteration (captured in the *Columns 1.01* software, to be described later) has a number of different methods programmed into it by which the settlement may be estimated. These are the methods proposed by Balaam and Booker (1981), Chow (1996), Bouassida et al. (2003), and the French Method (2005). In each case, the predicted settlement is compared to the adopted allowable settlement. The optimum value of IAR (η_{opt}) is captured when the predicted settlement is less than or equal to the specified allowable settlement. Users have the option of deciding which of the settlement prediction methods they prefer to adopt for this purpose. This decision will often be linked to the particular construction method adopted for the CRF (Bouassida and Carter, 2014).

4.4.2 Settlement of Floating Columns

Because it has been assumed that the soil behaves as a linear elastic continuum for the purpose of settlement analysis, the total settlement is the sum of the settlement component of the CRF soil mass, i.e., δ_r, and the settlement of the unreinforced underlying layer(s), δ_{ur}. The settlement component δ_r of the RS extending from the surface to a depth H_c is estimated using the method detailed previously for end-bearing columns.

However, for the case of floating columns, it is necessary to verify that the total settlement $\delta = \delta_r + \delta_{ur}$ is also within allowable limits. This verification is especially important when the unreinforced underlying layers are compressible saturated clays, in which case an estimate of the long-term consolidation settlement is also required. In such cases, the settlement of the unreinforced soil layers can be predicted by application of Terzaghi's theory of one-dimensional consolidation.

Further, a lower bound estimate of the settlement of the CRF may be derived using the theory of elasticity and, in particular, it may be expressed in terms of the *averaged* oedometer modulus, $E_{oed,r}$, of the reinforced ground (Bouassida et al., 2003a), defined by:

$$E_{oed,r} = (1-\eta)E_{oed} + \eta E_{oed,c} \tag{4.15}$$

The oedometer modulus has been previously defined by Equation (3.5b).

The oedometer condition provides the highest apparent Young's modulus of the CRF. In some practical cases, especially where the dimensions of the loaded area are much larger than the thickness of the reinforced layer(s), e.g., tanks having large diameters, estimation of the oedometer settlement may also be of interest, as suggested by Chow (1996).

The methodology of design detailed above has been recently incorporated in software known as *Columns 1.01*. The algorithm embodied in *Columns 1.01* is depicted schematically in Figure 4.1, and a summary of the optimization stage of the algorithm is illustrated in Figure 4.2 (Bouassida and Carter, 2014). This software and its algorithm have been successfully validated using numerous case histories. In particular, the effectiveness of the design procedure in terms of the eventual cost of the final project has been demonstrated by recorded field data for a tank project founded on stone columns (Bouassida and Hazzar, 2012).

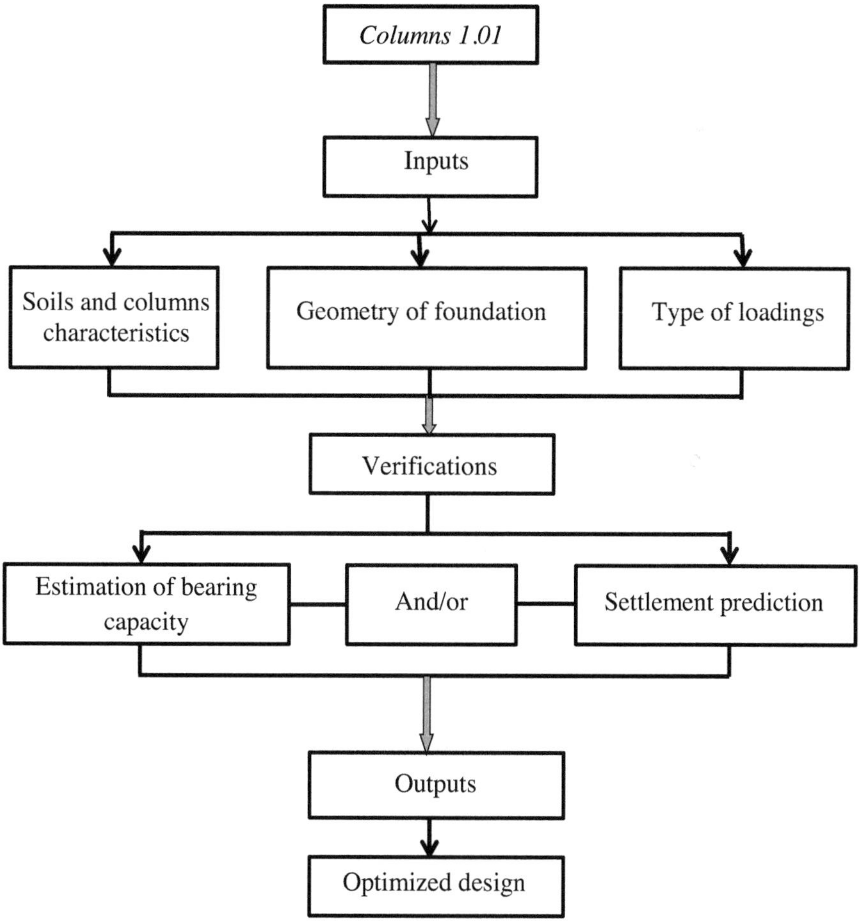

Figure 4.1 Algorithm of *Columns 1.01* software (Bouassida and Hazzar, 2012) (with permission of Institution of Civil Engineers)

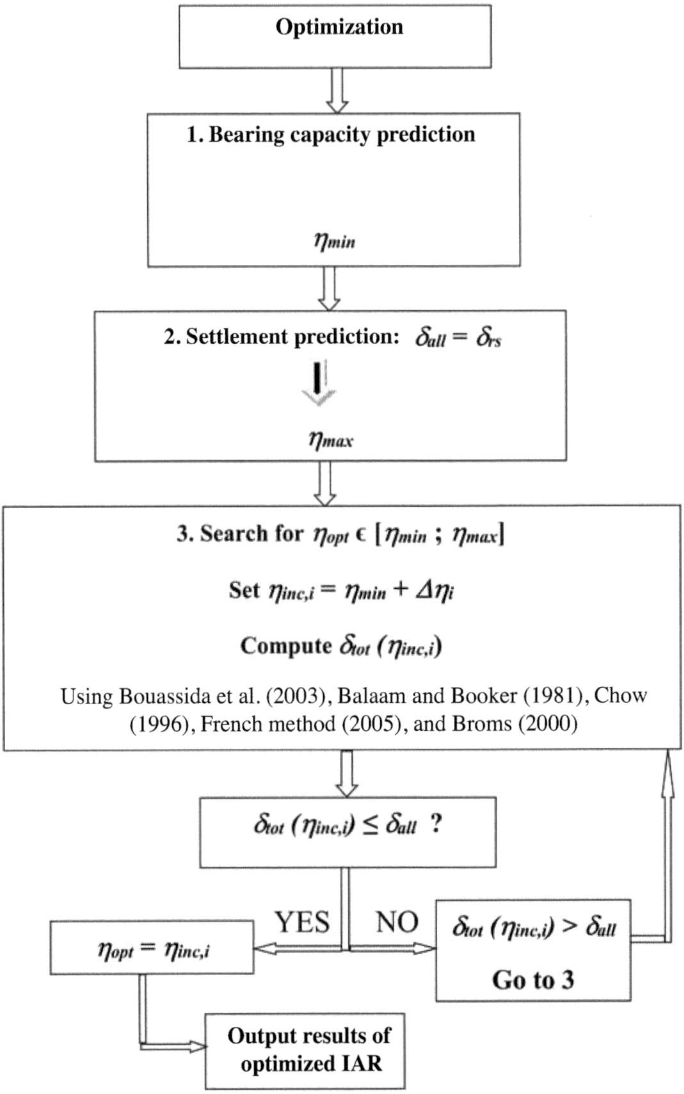

Figure 4.2 Optimization of IAR by *Columns 1.01* software (Bouassida and Carter, 2014) (with permission from ASCE)

4.4.3 Acceleration of Consolidation

When studying the reinforcement of soft clays by sand compaction columns and stone columns made of free-draining material having an enhanced permeability, the potential acceleration of the consolidation of the soft clay should be addressed in a second step, by considering the predicted optimized IAR, as previously detailed during the preliminary design of the CRF. A poro-elastic method has been implemented in *Columns 1.01* software to predict the evolution of accelerated settlement of the CRF, in such cases (Guetif-Fessi and Bouassida, 2005). This settlement, which depends in particular on the IAR, represents the long-term settlement of the RS layer, since the reinforcing columns also act as vertical drains.

It is also worth noting the improvement in the strength and deformation characteristics of the soft clay which results from its induced primary consolidation. Guetif et al. (2007) have reported results that indicate how Young's modulus of the soft clay is improved due to stone column installation and subsequent consolidation. As the stiffness of initial soil is improved in post stone columns installation, the shear strength of CRF is increased by consequence, even though the shear strength of columns material can be assumed unimproved. This provides a much better justification for the analysis of the long-term bearing capacity of the CRF which was initially predicted on the basis of the minimum IAR.

4.5 ILLUSTRATIVE CASE HISTORIES

4.5.1 Oil Tank on End-Bearing Sand Compaction Column Foundation (Tunisia)

An oil storage tank, 30 m in diameter, was built in the harbor area of La Goulette, a northern suburb of Tunis, Tunisia. The site was characterized by soft clay overlying firm sand. The tank loading was approximated by quasi-uniform vertical stress of 80 kPa, which does not exceed the allowable bearing capacity of the soft clay layer. Nevertheless, the predicted settlement of the tank of about 40 cm greatly exceeds the allowable settlement of 10 cm. Therefore, reinforcement by 10 m length end-bearing sand columns was adopted in order to reduce the final settlement and to accelerate significantly the consolidation settlement of the soft clay layer. Figure 4.3 summarizes the geotechnical properties of the host soil and the column material.

Using the *Columns 1.01* software, and adopting a maximum allowable settlement of the RS of 10 cm, the optimized IAR was estimated as 29.95 percent, which could be achieved by installing 729 sand piles, 0.6 m in diameter, with axis-to-axis spacing of 1.06 m in a triangular mesh. The corresponding predicted allowable bearing capacity of the CRF has been verified as having a safety factor greater than 2, which was considered reasonable for a tank project where the tank has a diameter of 30 m and the soft clay is only 10 m thick. Figure 4.4 shows the evolution of the normalized apparent Young's modulus of the CRF as

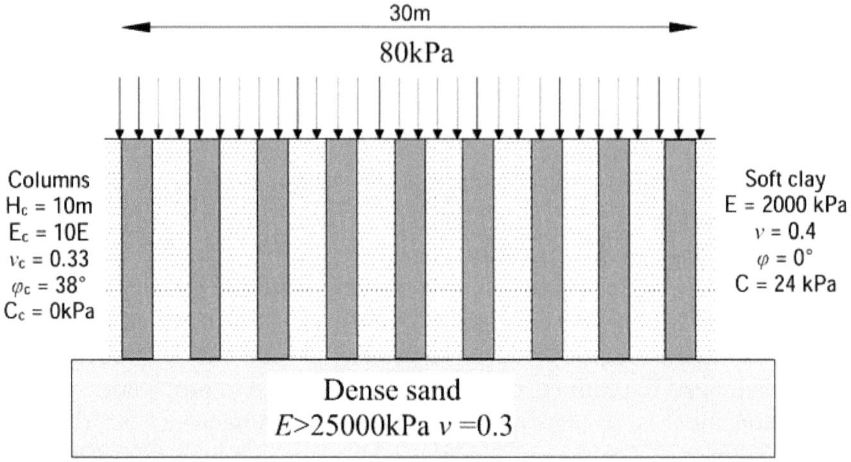

Figure 4.3 Circular tank on soft clay reinforced by end-bearing sand columns (with permission from ASCE)

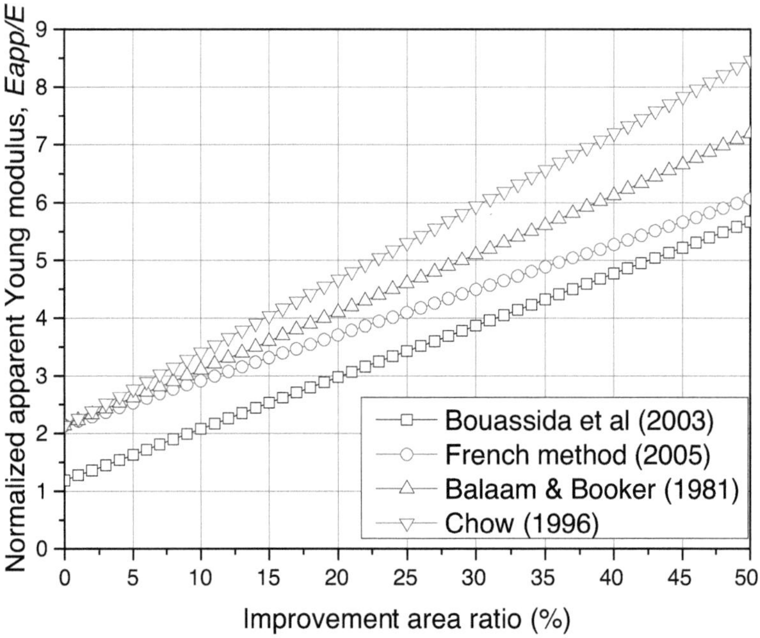

Figure 4.4 Predicted apparent Young's modulus of reinforced soil as a function of optimized IAR

a function of the IAR, from which it can be seen that the settlement predicted by the method of Bouassida et al. (2003) is conservative, probably due to the use of 3D modeling, without taking account of the improvement in the initial soil characteristics. By using the three other methods of settlement analysis, which are based on the UCM, with or without assuming the oedometer conditions, the settlement predictions are slightly underestimated. This settlement, which depends in particular on the IAR, represents the long-term settlement of the RS layer, since the reinforcing columns also act as vertical drains.

Figure 4.5 shows the evolution of settlement of the CRF as a function of the applied load for the optimized IAR. Note that the smallest settlements have been predicted by Chow's method because the horizontal component of displacement is neglected in this approach.

Furthermore, Figure 4.6 shows that the optimized area ratio, as predicted by the *Columns 1.01* software, increases from 14.6 to 36.3 percent when the

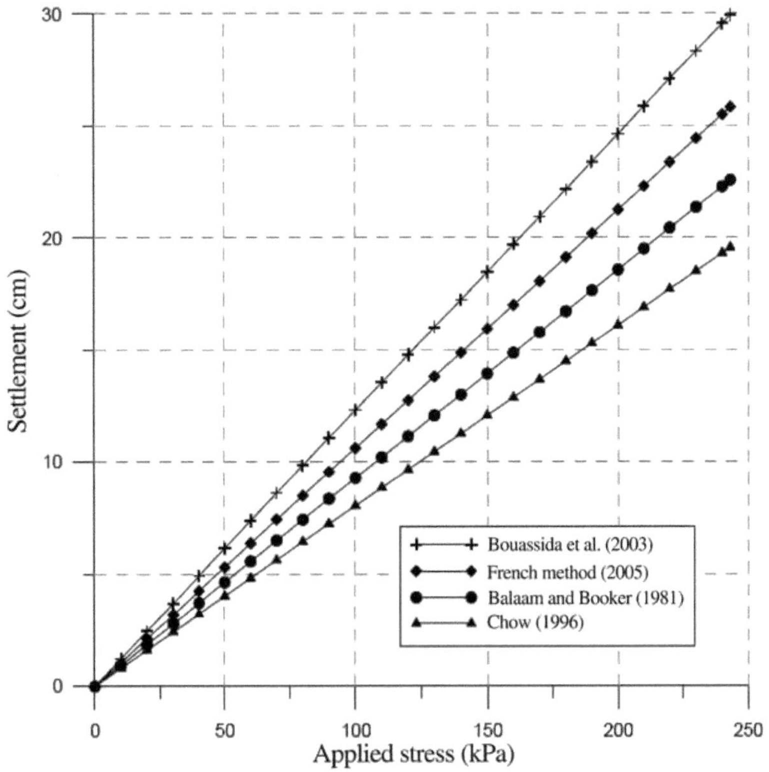

Figure 4.5 Predicted settlement as a function of the applied load for a tank on sand compaction columns (with permission from ASCE)

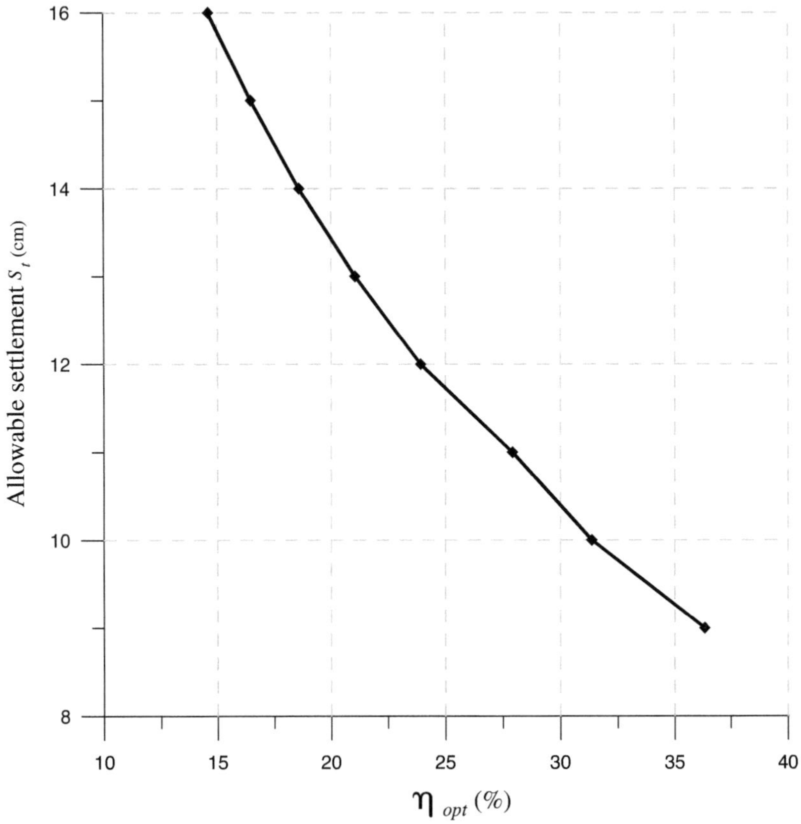

Figure 4.6 Evolution of the optimized IAR vs the allowable settlement of reinforced clay by end-bearing sand columns (with permission from ASCE)

allowable settlement of the column-reinforced foundation decreases from 16 to 9 cm. Such a case study illustrates that the main purpose of reinforcement by sand columns is to reduce settlement rather than to increase bearing capacity.

4.5.2 Embankment on Soft Clay Reinforced by Floating Columns

A road embankment, 2 m thick and 16 m wide at its base, was planned to be constructed on a soft clay layer that extends to 15 m depth beneath the ground surface (Figure 4.7). The soil profile is composed of two layers; the first was either cemented stabilized clay (CSC) or compacted sand fill (CSF) of a depth ranging from 1 m to 5 m. The second layer was natural soft clay, which extends to 15 m depth. In order to simulate different levels of live traffic loading, an

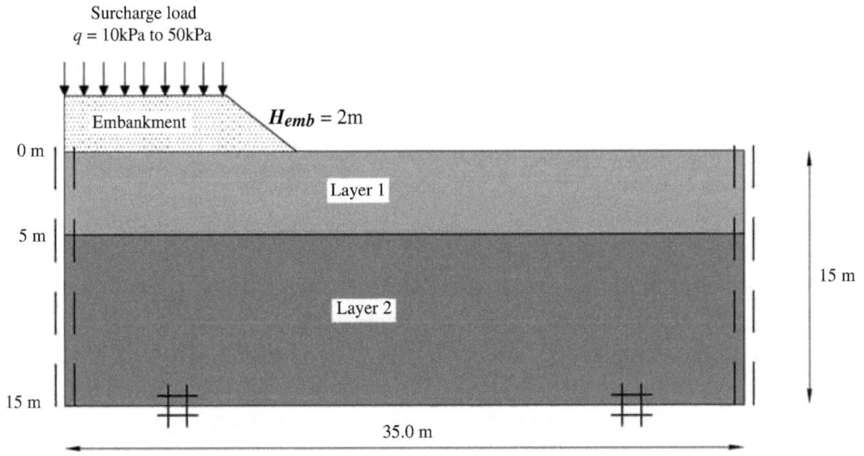

Figure 4.7 Numerical model studied by Saadeldine et al. (2011) (with permission from author Saadeldin, R.)

additional surcharge varying from 10 to 50 kPa was applied to the crest of the embankment. The weight of the embankment and the live loading were used in all settlement predictions.

The stability of the embankment was analyzed numerically to assess the performance of two proposed soil improvement techniques suggested by Saadeldin et al. (2011). The first technique would have involved treatment of the soft clay by continuous CSC of the upper layer, and the second technique was the substitution of the upper portion of the soft clay by CSF.

Instead of substituting the foundation layer beneath the embankment to a depth of 5 m by CSC or CSF, as assumed by Saadeldin et al. (2011), reinforcement of the soft clay by sand columns and columns formed by cement deep soil mixing has also been studied (Bouassida et al., 2012). The design procedure, incorporated in the *Columns 1.01* software, consisted of estimating the optimized IAR that complies with the allowable bearing capacity and settlement criteria for these latter two foundation options.

The optimized IAR depends on the maximum allowable settlement on the applied surcharge pressure and the depth of reinforcement. It was assumed that the allowable settlement equals 10 cm for the long-term embankment behavior. Once the optimized IAR is determined, the software computes, for different diameters, the number of columns to be installed and consequently the volume of reinforcing material to be added. It is then easy to compare the volume of material substituted in the scheme adopted by Saadeldin et al. (2011), i.e., η = 100%, to that predicted by the software. Tables 4.1 and 4.2 present the optimized IAR and the percentage of reduced material for compacted sand and CSC columns, respectively. It can be concluded that the reinforcement by columns is much more economical than the technique involving full substitution of weak soil to 5 m depth, as suggested by Saadeldin et al. (2011). Indeed, even if the length of columns exceeds 5 m, the optimized IAR would definitely be much less than that for total substitution ($\eta = 100\%$).

This second example illustrates again the efficiency of the software in estimating a truly cost-effective CRF.

Figure 4.8 represents a linear trend of the increase in surcharge embankment load as a function of the optimized IAR, as predicted by *Columns 1.01* software when half of the thickness of the soft clay layer is reinforced by DMM floating columns 7.5 m in length. The optimized IAR is predicted so that the settlement of the RS layer is equal to 10 cm. Hence, the linear variation from Figure 4.8 is expressed by (Bouassida and Carter, 2014):

$$q\ (\text{kPa}) = 40 + 0.9231\ (\eta_{opt}\ (\%) - 34.01) \tag{4.16}$$

Equation (4.16) indicates a direct relationship between the exerted load applied by the embankment and the optimized IAR.

Table 4.1 Optimized improvement area ratio for sand columns and percentage of savings of substituted sand fill

Surcharge (kPa)	Column length (m)	Optimized improvement area ratio (%)	(%) of savings of substitution material
20	7.5	17	75
30	7.5	31	54
40	7.5	44.5	33
50	7.5	58	12.5

Table 4.2 Optimized improvement area ratio for deep-mixed soil-cement columns and percentage of savings of substituted CSC

Surcharge (kPa)	Column length (m)	Optimized improvement area ratio (%)	(%) of savings of substitution material
20	7.5	56	15.5
30	7.5	60	10
40	8	31	53
50	8	31	50

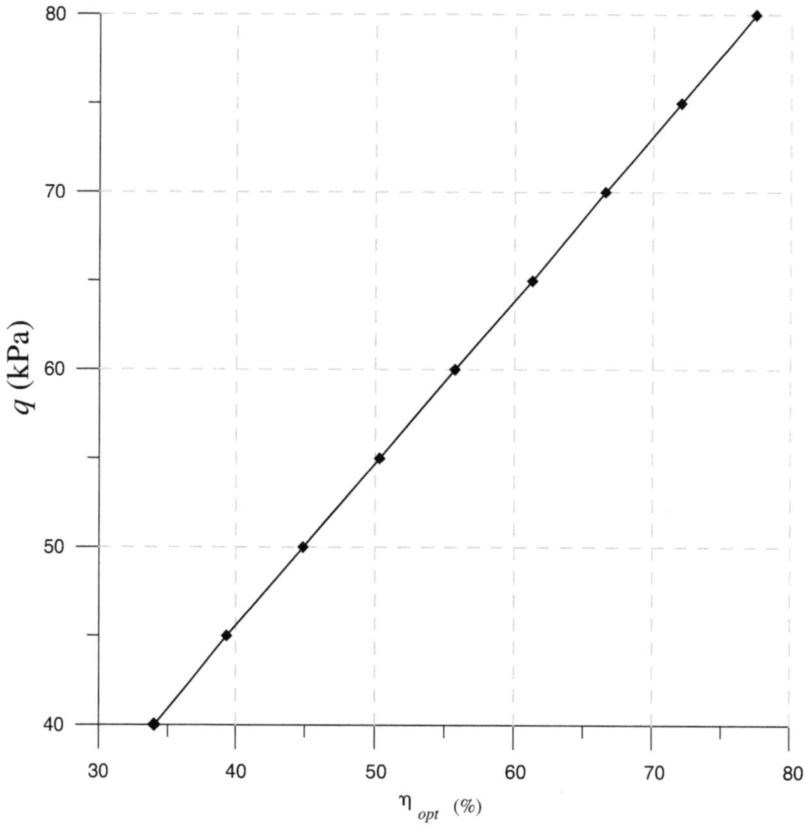

Figure 4.8 Evolution of load embankment versus the optimized IAR of soft clay reinforced by floating DMM columns of 7.5 m length in (with permission from ASCE)

4.5.3 Trial Embankment in Saga, Japan

An embankment 6.5 m in height was constructed on the soil profile illustrated in Figure 4.9, which shows floating DMM columns installed to a depth of 8.5 m (Chai and Carter, 2011).

This column reinforcement corresponds to a value of IAR = 30%, a value selected based on experience in other similar projects. The columns were installed within the soft Ariake clay layer, which is characterized by undrained shear strength of 12 kPa and a relatively high compression index of 2. Instrumentation for this trial embankment consisted of settlement recorders placed: (i) at the surface of the reinforced ground, i.e., points S_{01} and S_{02}, located respectively at the column head and on the original surface between adjacent columns, and (ii) in the RS mass at point S_1 and within the unreinforced soil layer at points S_2

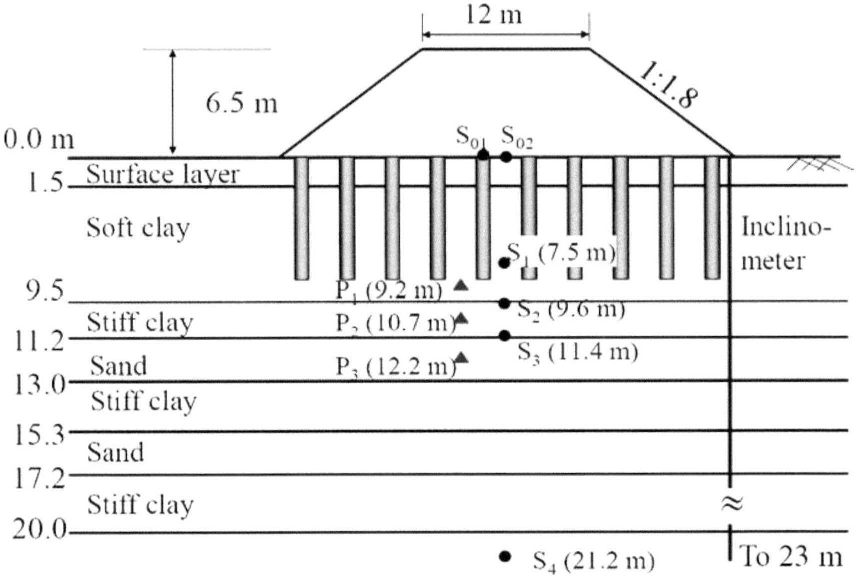

Figure 4.9 Embankment trial project in Saga, Japan (after Chai and Carter, 2011) (with permission from Springer)

and S_3. The evolution of the observed settlements is given in Figure 4.10, from which the following remarks are made:

1. At the surface of the reinforced ground, significant differential settlement of about 10 cm has been observed. This is essentially attributed to the big difference between the rigidities of the soft clay and column material. This behavior would not be expected if a thin mattress layer—at least 30 cm thick and made up of well-drained material—had been spread out over the ground surface in order to make the surface settlement of the reinforced ground as uniform as possible.
2. The long-term settlement of the unreinforced layers, about 19 cm in magnitude, is attained in a period of about 300 days. However, at the end of embankment construction, about 60 percent of this settlement (approximately 11 cm) had already occurred. The latter can be regarded as short-term settlement, corresponding to completion of the loading of the RS.

The Saga case history has been investigated by carrying out the complete design procedure incorporated in the *Columns 1.01* software (Bouassida and Hazzar, 2012), comprised of verifications of the bearing capacity and settlement. Table 4.3 summarizes the geotechnical properties of the soil layers adopted by Chai and

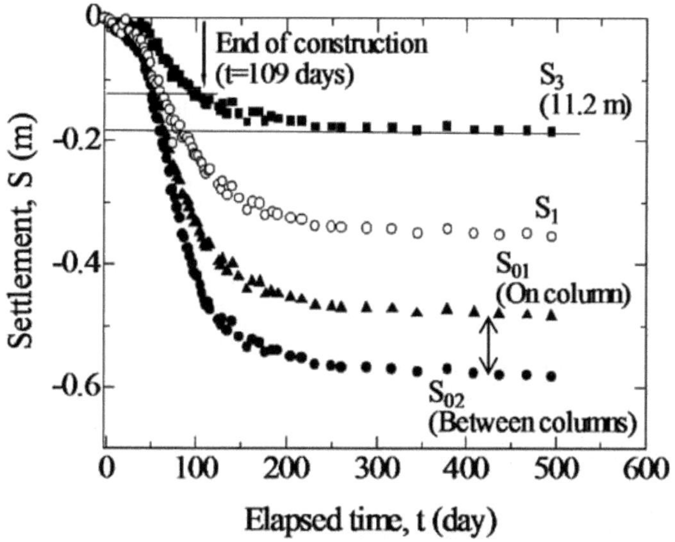

Figure 4.10 Observed settlements for Saga case history (Chai and Carter, 2011) (With permission from Springer)

Table 4.3 Geotechnical properties of soil layers at Saga site (with permission from ASCE)

Layer number	Thickness (m)	Undrained shear strength (kPa)	Young's modulus (kPa)	Poisson's ratio	Friction angle (degrees)	Unit weight (kN/m³)
1	1.5	15	2000	0.33	0	20
2	4.5	10	1500	0.45	0	13.5
3	3.5	20	3000	0.4	0	14.3
4	1.7	25	4000	0.35	0	18
5	1.8	0	15000	0.33	35	18
6	2.29	45	10000	0.35	0	18
7	1.9	0	17000	0.3	38	18.5
8	2.8	60	13500	0.33	0	19

Carter (2011), from which the characteristics of the column material have been deduced, i.e., $E_c = 30{,}000$ kPa, $v_c = 0.3$, $C_c = 300$ kPa, $\varphi_c = 0$ and $\gamma_c = 17$ kN/m^3.

Using these input data in the *Columns 1.01* software and assuming surface loading corresponding to an embankment having the dimensions given in Figure 4.9 and a unit weight of the embankment fill of 18.2 kN/m^3, the following predictions were obtained:

1. *Verification of bearing capacity*: It was found that a minimum IAR of 10.2% would be required to comply with an allowable bearing capacity corresponding to a safety factor of two.
2. *Verification of settlement*: Because floating columns were installed, this was achieved by comparing the total settlement, composed of the sum of the settlements of the RS and the underlying unreinforced soil layer, with the allowable settlement.

In addition, as observed in Figure 4.10, most of the settlement of the RS occurs at the end of construction. Accordingly, the optimized IAR has been sought in the range of total allowable settlement between 20 and 32 cm. From the settlement predictions obtained by the *Columns 1.01* software, the following interpretation has been made:

1. The total settlement of the unreinforced layers was predicted by the linear elastic model to be 12.5 cm. This prediction is quite similar to the recorded settlement at the end of construction of about 11 cm, as indicated in Figure 4.10 by the horizontal line (t = 109 days). It is inappropriate to compare this prediction of *Columns 1.01* with the observed consolidation settlement of the unreinforced layers, as the latter corresponds to long-term behavior.
2. It is also of interest to use *Columns 1.01* to predict how the overall settlement of the reinforced foundation system would vary as a function of the IAR (η) for the specified loading over the foundation, i.e., 6.5 m × 18.2 kN/m^3 = 118.3 kPa. Figure 4.11 shows the variation of the settlement as a function of the IAR, as predicted by the method of Balaam and Booker (1981). Plotted in Figure 4.11 are curves corresponding to the overall settlement of the foundation system, as well as that component of the overall settlement contributed by the layer of soil that is reinforced by columns. Obviously, the vertical separation of these two curves corresponds to the contribution to the overall settlement of the underlying unreinforced soil layer. Figure 4.11 indicates that the larger the allowable settlement of the RS, the smaller will be the required value of the IAR over the range of values of IAR > 10.22%. The lower limit on the IAR is required to satisfy the allowable bearing capacity criterion.
3. It would seem reasonable to consider that the average settlement within the RS mass at the end of embankment construction (as recorded at locations S_1 and S_{02}) is approximately 32 cm. It is noted that *Columns 1.01* predicts a linear elastic settlement, assumed to represent the short-term behavior of the RS, so that from Figure 4.11 it can be seen that a total allowable settlement of 32 cm corresponds to an optimized IAR

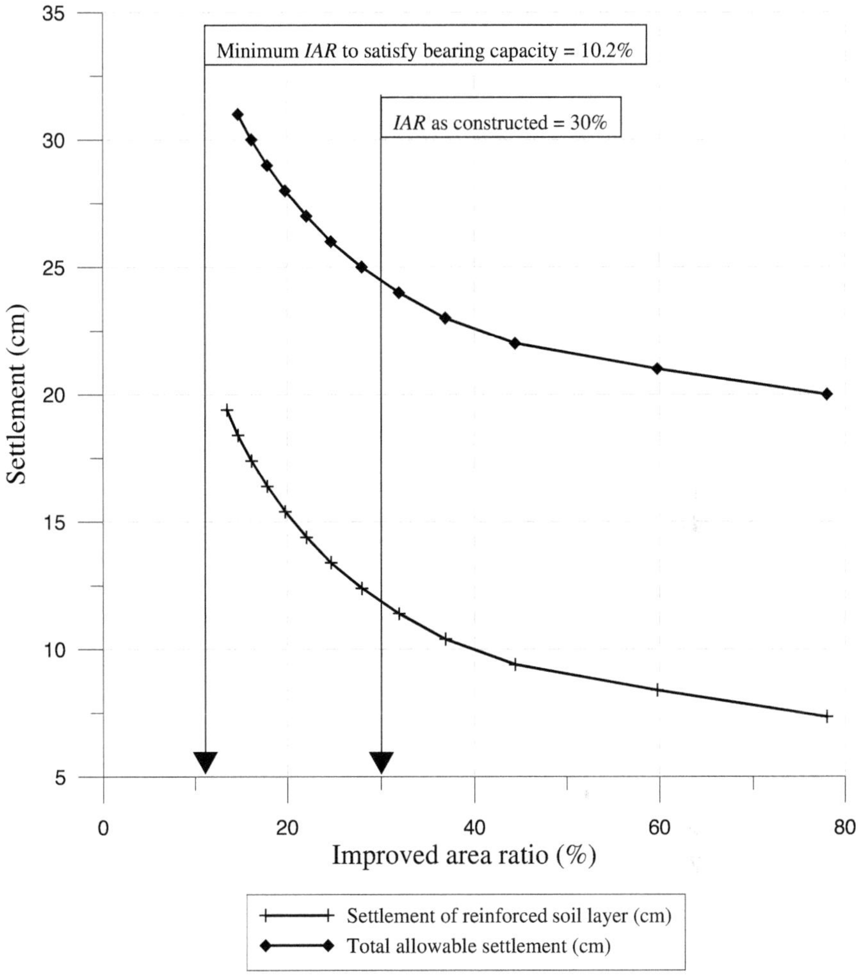

Figure 4.11 Settlements predicted by *Columns 1.01* for Saga case history as a function of the optimized IAR (with permission from ASCE)

of about 12%. Therefore, it can be concluded that there has been little benefit in constructing the columns with a value of IAR as large as 30%. The additional area (and volume) of column material corresponding to the higher value of the IAR was unnecessary for the settlement criterion to be met. Perhaps the main advantage of using a value of the IAR as large as 30% would appear to be as compensation for the non-use of a spreading layer of drained material at the ground surface to control the differential settlements.

From the three case histories analyzed and presented in this chapter it can be concluded that the *Columns 1.01* software provides a simple and efficient tool for predicting the optimized design of the CRF, both for end-bearing and floating columns.

4.6 CONCLUSIONS

A method has been suggested for carrying out the design of the CRF in soft ground. This comprehensive methodology successively considers the required bearing capacity and settlement criteria in the design procedure, in contrast to previous methods of design, which focus only on a unique criterion of design, i.e., settlement or bearing capacity alone. The latter methods are usually dedicated to some specific ground improvement technique. The main advantages of the suggested design method are its validity for all techniques of column installation and its applicability for both end-bearing and floating columns.

The IAR, which normally controls the cost of soil treatment, was targeted as a quantity to be optimized over a given range, satisfying the performance requirements of the foundation in terms of allowable bearing capacity and settlement. For practical purposes, the methodology has been implemented in the *Columns 1.01* software, which provides a viable tool for optimized and interactive design, being applicable to a variety of geotechnical structures.

Three different cases of study have been analyzed by the software. Predicted results demonstrate the efficiency of the proposed methodology in terms of cost effectiveness, specifically avoiding over-designed solutions.

Appendix 4.1

Reinforcement of a Purely Cohesive Soil with Cohesive-Frictional Column Material

Expressions for the ultimate vertical stresses acting on the column material and the host soil appearing in Equation (4.2), i.e., $\bar{Q}_{ult} / A = \sigma_{ult,rs} = (1-\eta)\sigma_{ult,s} + \eta\sigma_{ult,c}$, are given effectively as follows (after Bouassida et al., 1995):

For $1 \le K_{pc} \le 2$

Columns are located in an arbitrary pattern under the loaded foundation:

$$\left(\frac{\bar{Q}_{ult}}{A}\right)_{K_{pc}\le 2} = 4C(1-\eta) + 2\eta\left(K_{pc} + C_c\sqrt{K_{pc}}\right)$$

For $2 \le K_{pc} \le 1 + \left(\frac{b}{a}\right)^2$

Columns of identical radius are not necessarily located in a regular pattern under the loaded foundation. «*b*» is the minimum distance between axis-to-axis columns:

$$\left(\frac{Q_{ult}^{-}}{A}\right)=\left(\frac{Q_{ult}^{-}}{A}\right)_{K_{pc}\leq 2}+\eta C g(\varphi_c)$$

$$g(\varphi_c)=\left(K_{pc}-1\right)Ln(K_{pc}-1)-(K_{pc}-2)$$

The symbol a denotes the radius of the columns which are arranged in a regular pattern, for which the radius of the equivalent UCM is denoted by the symbol b.

The symbol $K_{pc}=tg^2(\frac{\pi}{4}+\frac{\varphi_c}{2})$ is used to denote the coefficient of passive earth pressure of the column material.

REFERENCES

Ambily, P. and Gandhi, S. R., Behavior of Stone Columns Based on Experimental and FEM Analysis. *J. Geotech. Geoenviron. Eng.*, ASCE, 133 (4), 2007, 405–414.

Balaam, N. P. and Booker, J. R., Analysis of rigid rafts supported by granular piles. *Int. J. of Num. & Analyt. Meth. in Geomech.*, 5 (4), 1981, 379–403.

Barksdale, R. D. and Bachus, R. C., Design and construction of stone columns. Final Report FHWA/RD 83-026. U.S. Federal Highways Administration, Washington, (1), 1983.

Bergado, D. T. and Lam, F. L., Full-scale load test on granular piles with different densities and different proportions of gravel and sands on soft Bangkok clay. *Soils and Foundations J.*, 27 (1), 1987, 86–93.

Bouassida, M. and Carter, J. P., Optimization of Design of Column-reinforced Foundations. *Int. J. Geomech.*, 14 (6) 2014, 10, 040143031-1-10.

Bouassida, M., Comprehensive design of column reinforced foundations. *Int. J. of Geotech. Eng.* Many & Sons Ltd, 7 (2), 2013, 156–164.

Bouassida, M., Hazzar, L. and Mejri, A., Assessment of software for the design of columnar reinforced soil. Proc. of Int. Symp. of ISSMGE - TC211. Recent research, advances and execution aspects of ground improvement works. Denies and Huybrechts (eds.) 31 May–1 June, 2012, Brussels, Belgium, III, 339–344.

Bouassida, M. and Hazzar, L., Novel tool for optimized design of reinforced soils by columns. *Ground Improvement*: Proc. ICE, London 165 (1), 2012, 31–40.

Bouassida, M., Jellali, B. and Porbaha, A., Limit Analysis of Rigid Foundations on Floating Columns. *Int. J. of Geomech.*, ASCE. 9 (3), 2009, 89–101.

Bouassida, M. and Porbaha, A., Ultimate bearing capacity of soft clays reinforced by a group of columns—Application to a deep mixing technique. *Soils and Foundations*, 44 (3) 2004, 91–101.

Bouassida, M., Guetif, Z., de Buhan, P. and Dormieux, L., Estimation par une approche variationnelle du tassement d'un sol renforcé par colonnes. *Revue Française de Géotechnique*, 102 (1), 2003, 21–29.

Bouassida, M., Guetif, Z., de Buhan, P. and Dormieux, L., Settlement estimation of a foundation resting on a soil reinforced by columns. Proc. 13th African Reg. Conf. on Soil Mech. and Geotech. Eng. Marrakech (Morocco) December 8–11, 2003a, 467–473.

Bouassida, M., Etude expérimentale du renforcement de la vase de Tunis par colonnes de sable—Application pour la validation de la résistance en compression théorique d'une cellule composite confinée. *Revue Française de Géotechnique*, 75 (2), 1996, 3–12.

Bouassida, M., de Buhan, P. and Dormieux, L., Bearing capacity of a foundation resting on a soil reinforced by a group of columns. *Géotechnique*, ICE, 45 (1), 1995, 25–34.

Broms, B. G., Lime and lime/columns. Summary and visions. Proc. 4th Int. Conf. on Ground Improvement Geosystems. Keynote Lecture. Helsinki, Finland, June 7–9, 2000, 43–93.

Chai, J. and Carter, J. P., *Deformation analysis in soft ground improvement.* Springer, 2011, Dordrecht.

Chow, Y. K., Settlement analysis of sand compaction piles. *Soils and Foundations J.*, 36 (1), 1996, 111–113.

El Ghabi, B., Hermges, T., Lambert, S., Ellouze, S., Bouassida, M. and Mejri, A., Comparison between predicted settlement of a group of stone columns foundation and in situ records. Proc. 2nd Int. Conf. on Geotech. Eng., Bouassida et al. (Eds), October 25–27, 2010, Hammamet (Tunisia), 255–264.

Ellouze, S., Bouassida, M., Hazzar, L. and Mroueh, H., On settlement of stone column foundation by Priebe's method. *Proc. of the Institution of Civil Engineers Ground Improvement J.*, 163 (2) 2010, 101–107.

French Method. Recommandations sur la conception, le calcul, l'exécution et le contrôle des colonnes ballastées sous bâtiments et ouvrages sensibles au tassement. *Revue Française de Géotechnique*, 11 (2), 2005, 3–16.

Ghionna, V. and Jamiolkowski, M., Colonne di ghiaia. X Ciclo di conferenze dedicate ai problemi dei terreni e ingegneria delle fondazioni metodi di miglioramento dei terreni, Politecnico di Torino Ingegneria, atti dell'istituto di scienza delle costruzioni, 507, Nov. 1981.

Guetif Fessi, Z. and Bouassida, M., Analytical estimate of settlement evolution of soft soil reinforced by stone columns. Proc. 16th Int. Conf. of Soil Mech. and Geotech. Eng. September 12th-16th 2005, Osaka, Japan, 3:1355–1358.

Guetif, Z., Bouassida, M. and Debats, J. M., Improved soft clay characteristics due to stone column installation. *Computers and Geotechnics*, 34 (2), 2007, 104–111.

Kitazume, M., Ikeda, T., Miyajima, S. and Karastanev, D., Bearing capacity of improved ground with columns type DMM. Proc. IS-Tokyo '96. 2nd Int. Conf. on Ground Improvement Geosystems, 1996, 503–508.

Poorooshasb, H. B. and Meyerhof, G. G., Analysis of behavior of stone columns and lime columns. *Computers and Geotechnics*, 1997, 1 (20), 47–70.

Priebe, H. J., The design of vibro replacement. *Ground Engineering* 28 (10), 1995, 31–37.

Saadeldin, R., Salem, M. A. and Lotfi, H. A., Performance of road embankment on cement stabilized soft clay. Proc. 14th Pan-American and 64th Canadian Geotech. Conf., October 2–6, 2011, Toronto, Ontario, Canada.

5

THE BEHAVIOR OF IMPROVED SOFT CLAY BY STONE COLUMN INSTALLATION

ABSTRACT

This chapter deals with the installation effects of stone columns in soft soils. The first part focuses on the lateral expansion of stone material using the vibro-displacement and substitution techniques by means of numerical simulations and laboratory tests. The second part exposes a study on the behavior of a reinforced soil after stone column installation to show how the properties of soft soils can be improved prior to final loading. The effect of such an improvement on the prediction of the settlement of reinforced soil is then evaluated. Two case studies are discussed to highlight the installation effects of stone columns in soft soils.

The experimental behavior of remolded kaolin clay reinforced by stone columns is then investigated. The installation of stone columns was simulated in the laboratory by performing a lateral expansion within hollow cylindrical remolded kaolin specimens. The granular material is incorporated in the hollow cavity to form the stone column. After column installation, specimens were subjected to consolidated undrained (CU) triaxial shear tests with recorded excess pore pressure. The effects of consolidation stress and the installation of a stone column were analyzed based on the improvement of mechanical properties of kaolin clay, i.e., undrained Young's modulus and undrained cohesion.

Finally, a linear poroelastic model is formulated to derive the development of primary consolidation settlement by drained column material for a given history of loading (stage construction).

5.1 INTRODUCTION

Reinforcement of soft clay by stone columns has become, over the last three decades, one of the most practiced soil improvement techniques in civil engineering projects. Vibro-displacement and vibro-substitution techniques made the installation of stone columns possible by inducing horizontal vibrations in soft soil. The stone material, by means of a vibro-probe, is laterally expanded in the soft soil in which compression depends on the consistency of the crossed layers (Figure 5.1). It is important to note that the cross section of the installed column also depends on soil consistency. Once the column is installed, its cross

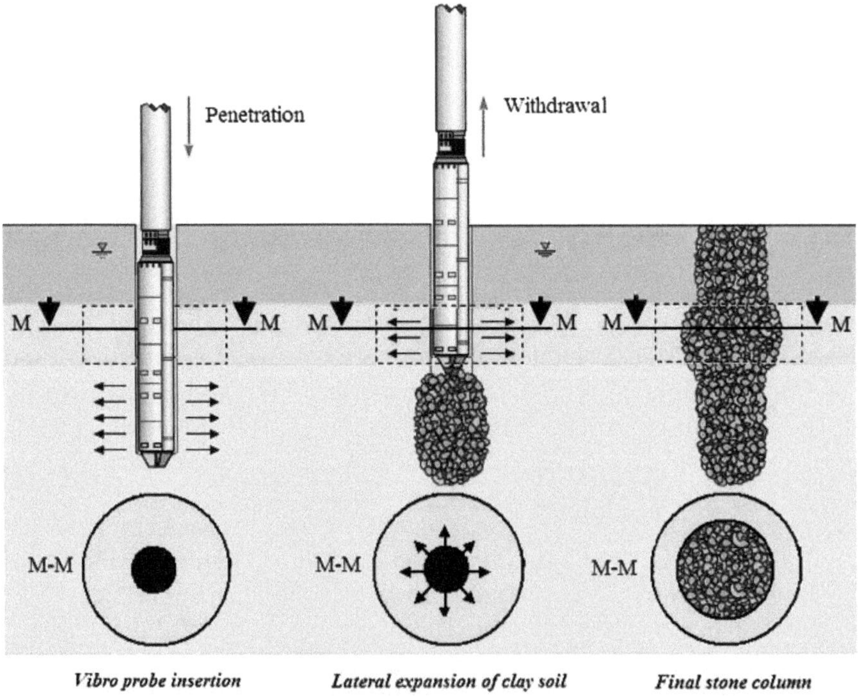

Figure 5.1 Schematic of vibro-replacement installation of stone column by the dry method (modified source from www.keller-france.com)

section is not uniform, but an equivalent area can be calculated by measurements of the volume of incorporated stone material and depth of treatment. Lateral expansion of the stone material represents a similar loading to that applied during the pressuremeter test, which induces excess pore pressure in the soft soil. As a result of the enhanced drainage property of the column material, a primary consolidation is triggered within the soft clay. Due to the reduced drainage path between the installed stone columns (less than their axis-to-axis spacing), the consolidation of soft clay mostly happens horizontally, resulting in the dissipation of excess pore pressure that is accompanied by an increase in effective stresses (Debats et al., 2003).

In situ control tests that were carried out before and after the installation of stone columns have given evidence of the improvement of the existing soft soil properties (Figure 5.2).

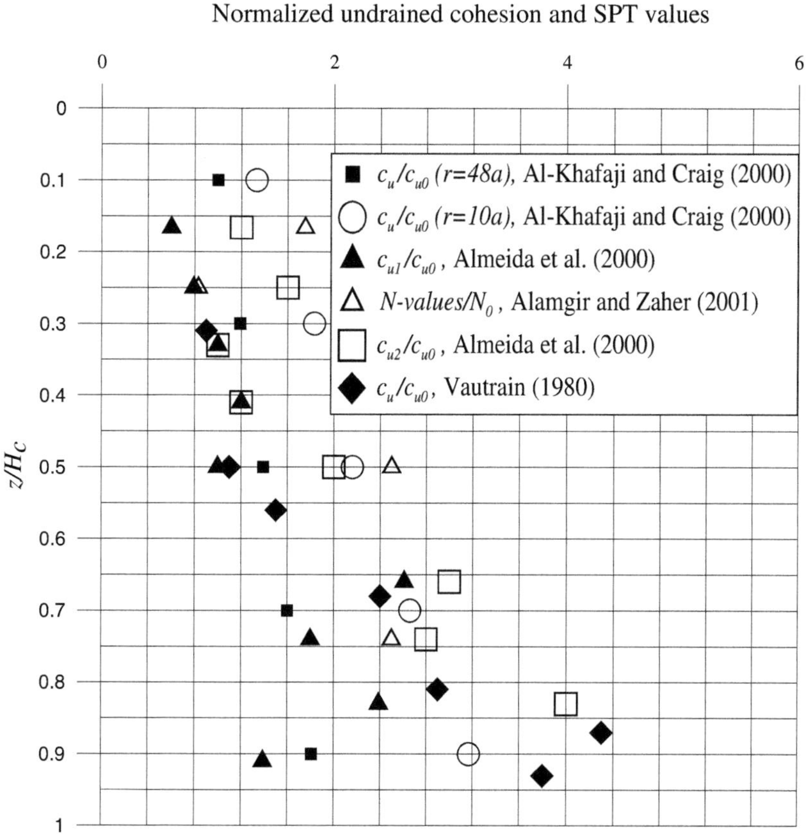

Figure 5.2 Normalized undrained cohesion and SPT values in improved soft soils by stone columns

Predictions of improved properties (Young's modulus and undrained shear strength) of soft soil as well as the extent of the improved zone surrounding the stone column are extremely important for predicting the bearing capacity and settlement of the reinforced soil. Indeed, the post final loading behavior of reinforced stone column foundations depends on the improved properties of the soft soil, especially for settlement prediction (Ellouze et al., 2010).

Further, because of the enhanced drainage property of stone material, the columns play the role of vertical drains. Hence, the settlement of reinforced soil will be mainly accelerated by horizontal drainage. The development of settlement, in particular when stage construction is planned, gives the best insight about the behavior of stone column-reinforced foundations (CRF).

This chapter studies the behavior of reinforced soft clay from the installation of a stone column, as a first stage, up to the acceleration of consolidation associated with the development of settlement of the reinforced soil when subjected to its final loading.

Section 5.2 presents the installation of a stone column by lateral expansion and the induced improvement of mechanical properties of the soft clay prior to the final loading.

The experimental work and numerical simulations pertaining to such installation are detailed in Section 5.3. Predictions obtained by these simulations are discussed and then compared.

Section 5.4 is dedicated to the study of settlement reduction by taking into account soft clay improvement resulting from the installation of stone columns by lateral expansion. Recommendations for the design of a stone column foundation are then suggested.

In Section 5.5, the acceleration of settlement of stone CRF subjected to final loading is investigated by means of a poroelastic model proposed by Fessi Guetif (2004) and Guetif Fessi and Bouassida (2005).

5.2 STONE COLUMN INSTALLATION IN SOFT CLAY

The vibro-installation of stone columns was designed so that the improvement of soft soil takes place at a prescribed depth, by the lateral expansion of added stone material. This expansion is induced by horizontal vibrations powered by an eccentric motor within a vibro-probe. Figure 5.1 illustrates that the diameter of installed stone columns by the dry (vibro-displacement) method depends on the consistency of the crossed soft soil layer. Lateral expansion represents a similar loading to that produced when a pressuremeter test is carried out (Frikha et al., 2008). Hence, because the stone column is installed rapidly, excess pore pressure is induced in the surrounding soft soil from which a primary consolidation is triggered progressively as a result of the horizontal drainage toward the installed stone column.

The observed behavior of soft soil improved by vibro-stone columns has been extensively investigated from measurements during in situ control tests that were executed during the pre- and post-treatment phases. For instance, Vautrain (1980) and Bergado et al. (1984) have reported an increase in undrained shear strength (c_u) of the soft soil in post-columns installation from several vane shear test results. For Bangkok soft clay, the increase in undrained cohesion was about 50 percent after the installation of stone columns (Bergado et al., 1984). Gambin (1984), Callanan (1991), Kirsch (2005), and Corneille et al. (2006) have confirmed the increase, at different proportions, in the resistance of soft soil due to the installation of stone columns by performing different series of standard penetration tests (SPTs), cone penetration tests (CPTs), and pressuremeter tests.

Figure 5.2 shows the variation of normalized undrained cohesion and the SPT number as a function of normalized depth in post-treatment of soft soils by stone columns with respect to the pre-treatment phase. It is shown that significant improvement of soft soil characteristics has occurred as a result of stone column installation by lateral expansion (Guetif et al., 2007).

Debats et al. (2000) reported an interesting French case history (a project of the Mediterranean line of a high-speed train) dealing with the comparison between the effects of static preloading and the improvement by means of the vibro-replacement method. Such improvement was dictated due to the presence of fine silt sand layers to minimize the risk of liquefaction that could be induced by seismic loading. The control tests included classical cone penetration and Menard pressuremeter tests carried out within two zones. The first zone corresponds to the vibro-replacement improved soil with an area equal to 10.6 m^2. The second zone is near the abutments of a bridge (founded on end-bearing piles) and was subjected to a static preloading embankment of 9 m in height.

Figures 5.3 and 5.4 compare the measurements of tip resistance obtained from a CPT within and outside the preloaded zone, respectively. Those figures clearly show that an at-rest period of four months between the end of treatment and the time of control tests is beneficial. During this at-rest time, a significant degree of improvement occurred between 2.5–5.5 m in depth. Figures 5.5 and 5.6a indicate that similar degrees of improvement were obtained for the tip resistance and Menard's limit pressure. The Menard's modulus seems to be, for finer soils, a better indicator of soil improvement due to the installation of a stone column (Figure 5.6b). Investigation of the French case history confirmed that the choice of control test types was of considerable interest to accurately estimate the improvement characteristics of fine granular soils after treatment by the vibro-substitution method.

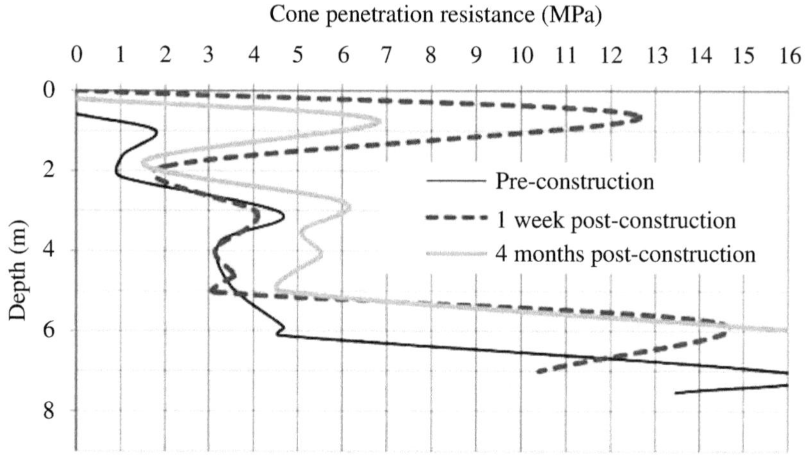

Figure 5.3 Cone penetration resistance recorded within the preloaded zone (with permission from author Gambin, M.)

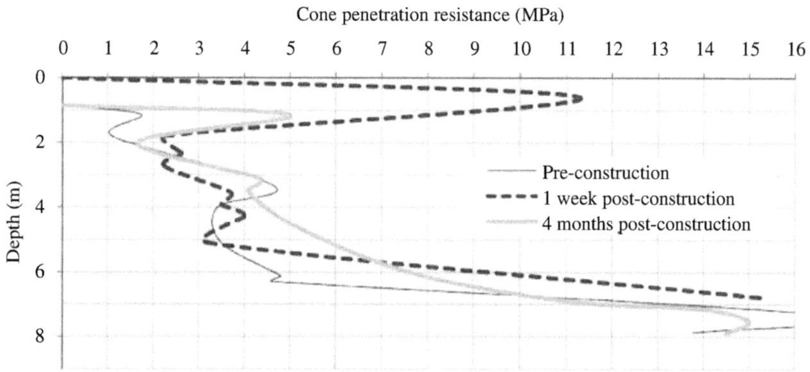

Figure 5.4 Cone penetration resistance recorded outside the preloaded zone (with permission from author Gambin, M.)

A reliable study of the behavior of soft soils in post-stone column installation needs the collection of numerous measured data from comparable case histories, which is almost impossible to perform. Therefore, numerical simulation techniques of the installation of stone columns in soft clays seem to be more appropriate to explain the behavior of improved soft soil.

Nevertheless, performing laboratory experiments on scaled test models of reinforced soil by stone columns also represents an interesting approach to quantify the degree of intended improvement. In the following section, numerical simulations of the installation of stone columns in soft soils are described.

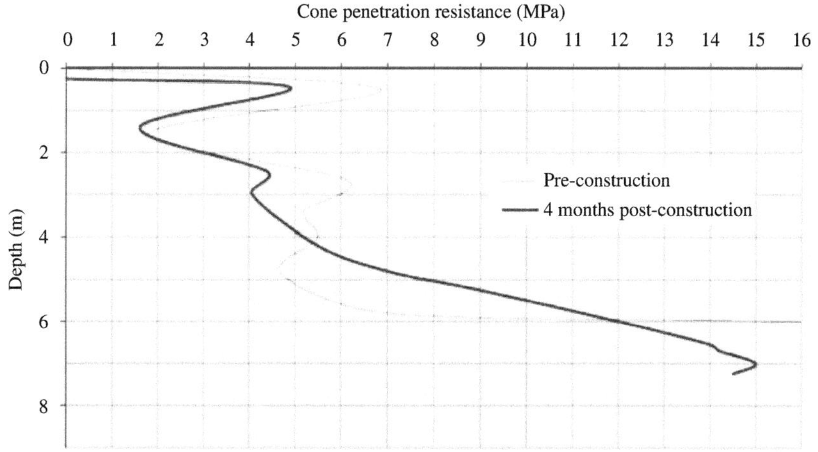

Figure 5.5 Compared cone penetration resistance recorded four months in post-column installation (with permission from author Gambin, M.)

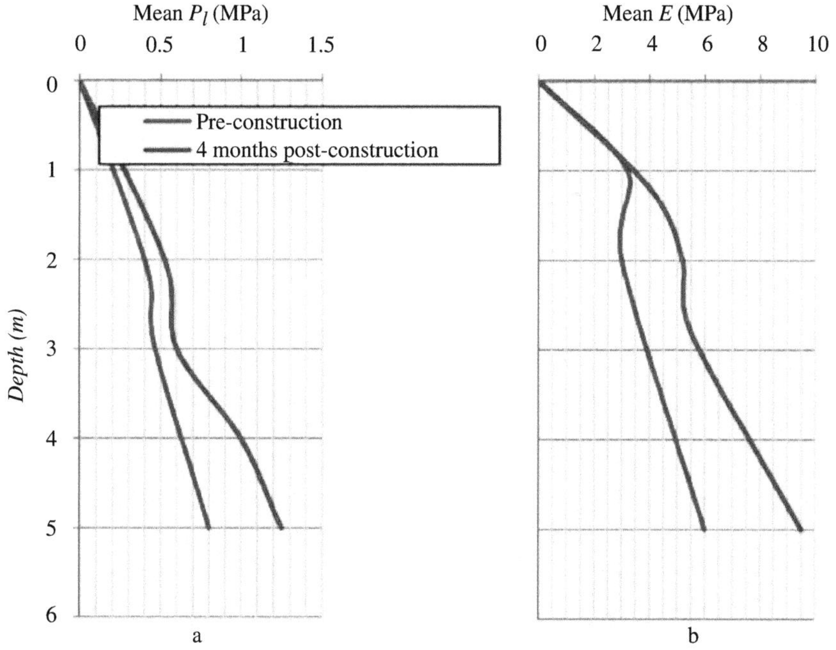

Figure 5.6 Compared pressuremeter recorded parameters before and after column installation—4 months in post treatment phase (with permission from author Gambin, M.)

5.3 SIMULATION OF STONE COLUMN INSTALLATION IN SOFT SOILS

Studies about soft soil improvement resulting from stone column installation were initiated about a decade ago through a collaborative research work between the National Engineering School of Tunis (The University of Tunis El Manar) and the Vibroflotation Group (Aix en Provence, France). This collaboration was part of a development program dedicated to Plaxis software on which several European research teams were involved. In this regard, contributions made earlier by Debats et al. (2003), followed by Guetif et al. (2007), Ellouze and Bouassida (2009), Ellouze Zribi (2010), and by Bouassida et al. (2012) represent a milestone for the numerical prediction of improved soft clay properties due to stone column installation.

5.3.1 Numerical Simulations Using the Unit Cell Model

Installation of vibro-compacted stone columns within soft clay was simulated using Plaxis software in axisymmetric condition. Prediction of soft soil improvement was based on the change of its Young's modulus between the phases of pre-treatment and post-treatment followed by the inherent primary consolidation induced by the installation of stone columns. This adopted method of estimation is mainly based on the correlation proposed by Biarez et al. (1998) from triaxial test data that gives the change of Young's modulus as a function of the effective mean stress:

$$\frac{E}{E_{ref}} = \left(\frac{\sigma'_m}{\sigma'_{mref}} \right)^{\varpi} \tag{5.1}$$

$\sigma'_m = (\sigma'_1 + 2\sigma'_2)/3$ and σ'_{mref} denote the current effective mean stress and its value for a given state of reference. Similarly, E denotes the current Young's modulus and E_{ref} denotes its value at the same state of reference.

The exponent ϖ appearing in Equation (5.1) is a dimensionless empirical parameter that predicts the change of the normalized Young's modulus as a function of the normalized effective mean stress. For normally consolidated soft clays, the Plaxis user's manual recommended a value of 1 for ϖ (Brinkgreve and Vermeer, 1998). This means that the change of effective mean stress equals that of Young's modulus. In a first stage, the perfect Mohr-Coulomb's elastoplastic behavior was recommended, the Young's modulus of the soft clay is estimated from its secant Young's modulus E_{50} that corresponds to half of the ultimate deviatoric stress sketched in the stress-strain curve obtained from a triaxial shear test (Figure 5.7). Note that the deviatoric stress is:

$$q = \sigma'_1 - \sigma'_3 \tag{5.2}$$

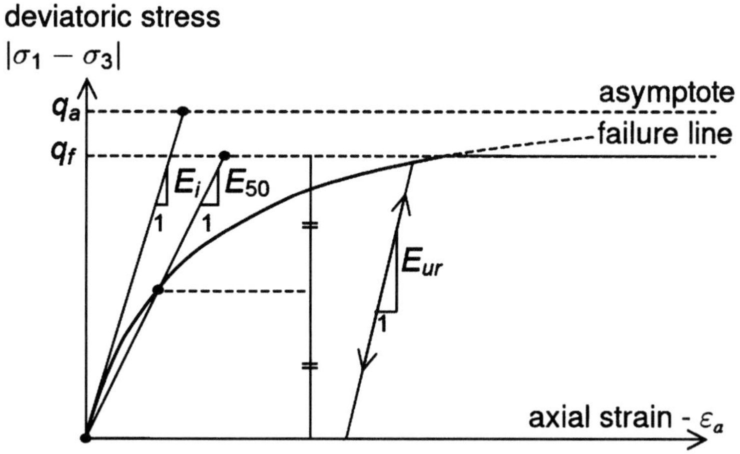

Figure 5.7 Determination of Young's moduli of soil modeled by perfect elasto-plastic behavior (Mohr Coulomb model). (With permission from Plaxis BV.)

Therefore, correlation given by Equation (5.1) was adopted to predict the change in Young's modulus from a given reference state to the current state of stress-strain within the soft clay. In the following, the initial state, as referred to the state of reference, is denoted by a subscript zero "$_0$" and corresponds to the end of stone column installation within the soft clay.

It was suggested by the Plaxis team to simulate the lateral expansion of a stone column within the soft clay according to the steps shown in Figure 5.8. Debats et al. (2003) reported the primary implementation of the procedure of dummy material to predict the behavior of soft clay in post installation of a vibro-compacted stone column.

The dummy material procedure has been performed for two case studies of stone column installation in soft clay. Interface between the columns and the soft clay is assumed to be perfect contact (total adhesion). The first case study considers the reinforcement of soft clay by an end-bearing stone column that is modeled by a unit cell model (UCM). Mohr-Coulomb and hardening soil constitutive laws were adopted to compare the behavior of soft clay predicted in pre- and post-installation of a stone column.

The second case study deals with the analysis of reinforcement by a group of stone columns. The Mohr-Coulomb's constitutive law is then considered to compare predictions of the settlement reduction due the installation of stone columns.

For all numerical computations, the improved soil model is discretized by means of 15 nodes triangular finite elements by using the automatically generated mesh option provided by Plaxis software. The Mohr-Coulomb's constitutive law is adopted for constituents of the reinforced soil model. The state of stress at

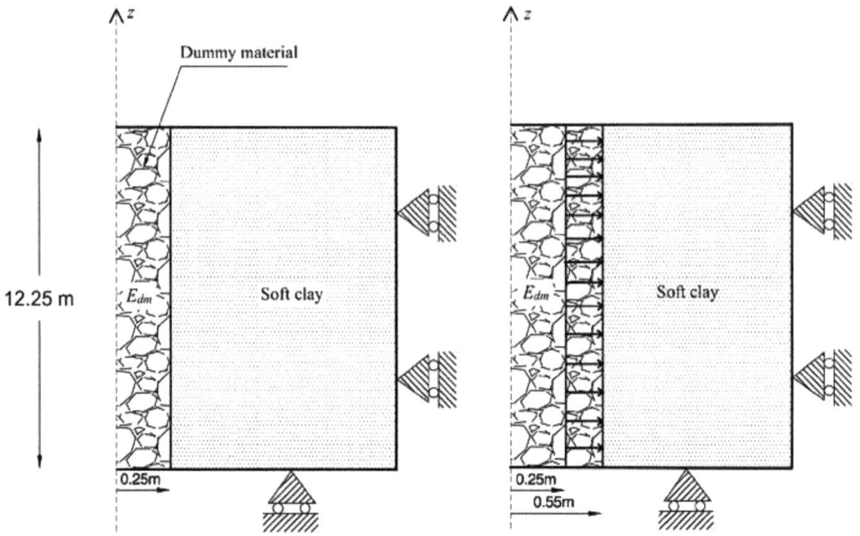

Figure 5.8 Simulation of stone column installation in a soft clay layer (with permission from author Guetif, Z.)

rest has been adopted as the initial condition, both in the columns and in the soft clay, using the at-rest earth pressure coefficient: $K_0 = 1 - \sin \varphi$.

Another case study was considered for the Damiette's project investigated by Namaa Engineering & Consulting (2001), with a soil profile made up of a soft clay layer of 12.25 m thickness overlaid by a 12.75 m-thick sand layer (Figure 5.9). The soft clay layer was reinforced by vibro-compacted stone columns with a diameter of 1.1 m using a triangular mesh of 2.7 m spacing. The improvement area ratio (IAR) is 15 percent. The upper sand layer is vibro-compacted without added material to increase its relative density.

The effect of the soft clay improvement due to the installation of the stone column was studied by performing numerical analysis using Plaxis version 9.2D. An axisymmetric UCM (as shown in Figure 5.9) with an external radius of 5 m was adopted for that study (Guetif et al., 2003).

The dummy material procedure is implemented according to the following steps:

- Penetration of the vibro-probe within the soft clay layer is simulated by the creation of an initial hole, 0.25 m in radius, modeled as a linear elastic fictitious material having a very weak Young's modulus $E_{dm} = 20kPa$ (Figure 5.8).
- Lateral expansion of stone material within the soft clay layer is simulated as an imposed uniform horizontal displacement of 0.30 m applied at the border of the initial hole to reach the final radius of the stone column

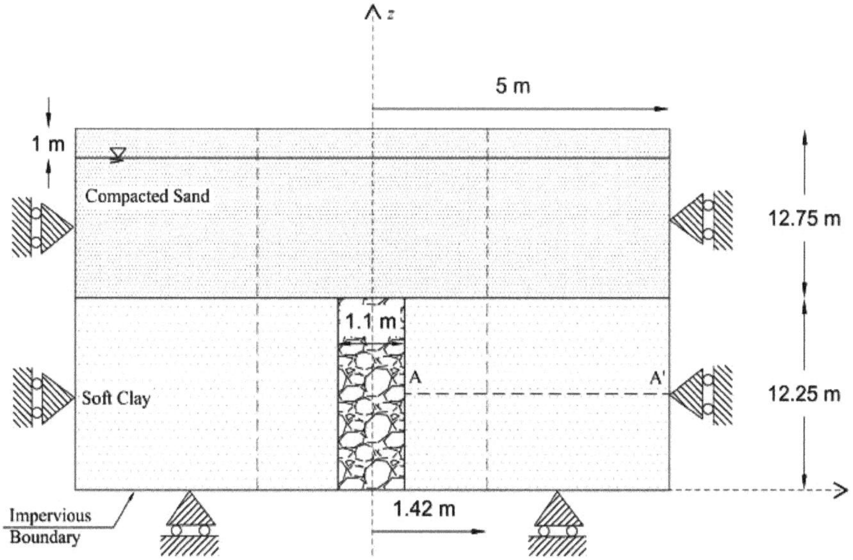

Figure 5.9 Axisymmetric UCM of reinforced soft clay by a vibro-compacted stone column and sand layer (with permission from author Guetif, Z.)

of 0.55 m (Figure 5.8). Then, the fictitious material is substituted by the stone material in terms of geotechnical properties.

5.3.1.1 Numerical Predictions by the Mohr Coulomb Model

The Mohr Coulomb constitutive law is adopted for the constituents of the UCM (Table 5.1). These parameters have been adopted on the basis of data found in the geotechnical report of Damiette's project. The drained Poisson's ratio for all materials was taken equal to 0.33.

Figure 5.10 displays the deformed mesh within the soft clay layer in the post-stone column installation phase.

Table 5.1 Natural characteristics of reinforced soil by a group of end-bearing columns, Damiette's project (Naama Engineering & Consulting, 2001)

Material	Total unit weight (kN/m^3)	Drained Young's modulus (kN/m^2)	Drained cohesion (kN/m^2)	Drained friction angle (°)	Isotropic permeability (m/day)
Soft clay	17	4000	5	21	0.94 10^{-5}
Column material	20	32000	10	38	100
Initial sand	18	25000	1	35	10
Compacted sand	20	50000	1	38	10

A significant excess pore pressure occurred (u_0 = 70 to 100 kPa) immediately after the installation of a stone column within the soft clay layer. An increase in horizontal effective stress of about 50% is predicted with a decrease in vertical effective stress. However, total effective mean stress remains almost constant. This led to a conclusion that improvement in Young's modulus of soft clay cannot occur immediately after the stone column installation.

The simulated immediate post-behavior, after stone column installation, is illustrated by Figure 5.11. This figure shows, at mid-depth of the soft clay layer, the variation of normalized excess pore pressure as a function of horizontal distance from the border of the stone column to the border of the UCM. Numerical predictions within a stone column showed a significant increase in effective stresses (3.6 to 5.6 for a horizontal component and 2.2 to 2.7 for a vertical component). This explains the increase in effective horizontal stresses which has resulted from the lateral expansion of column material in soft clay.

Figure 5.10 Deformed mesh in soft clay layer in post installation of stone column (Mohr Coulomb model) (with permission from the University of Thu Dau Mot, editors Hai et al., 2013)

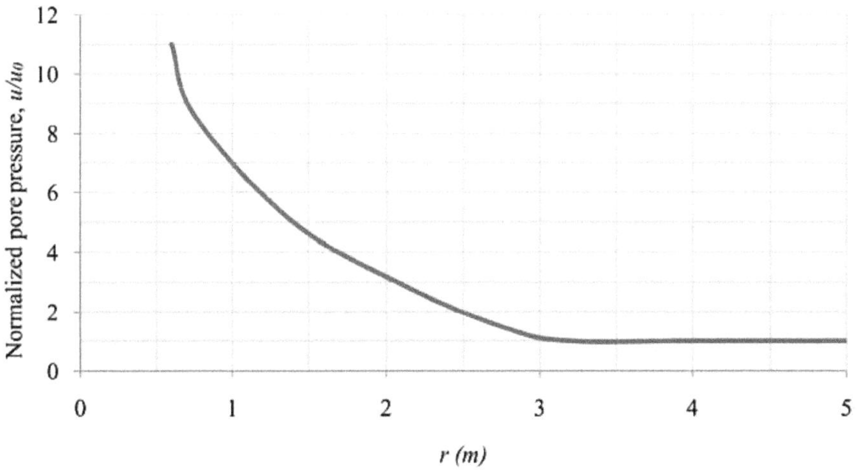

Figure 5.11 Variation of normalized excess pore pressure in soft clay vs. distance from stone column immediately after stone column installation (with permission from the University of Thu Dau Mot, editors Hai et al., 2013)

After column installation, the primary consolidation option embodied in Plaxis's menu was run for a period of eleven months. Parametric study showed that the influence of boundary conditions does not affect numerical predictions when the external radius of the UCM is equal to 5 m (Fessi-Guetif, 2004).

After eleven months of consolidation in soft clay, predictions are: total dissipation of excess pore pressure in soft clay happens with a significant increase in horizontal effective stress. In turn, the vertical effective stress remains constant.

Figure 5.12 shows that a significant increase (by 80%) in Young's modulus occurs close to the stone column edge. However, it abruptly vanishes at a distance equal to four times the stone column radius (4 × 0.55 m = 2.2 m).

Numerical predictions indicate that Young's modulus of the soft clay increases by approximately 32%. Such an increase resulted from eleven months of primary consolidation after the stone column installation. Further, Figure 5.12 shows that the influence zone radius of improved soft clay is nearly four times the final radius of the stone column. Hence, it can be concluded that the settlement reduction of reinforced soil is due not only to the reinforcement role of the stone column made up of a stiffer material, but also to the improvement in Young's modulus of the soft soil.

5.3.1.2 Numerical Predictions by the Hardening Soil Model

In a prior work, Guetif et al. (2008) investigated the hardening soil model (HSM) for constituents of the UCM shown in Figure 5.9. The parameters of the HSM

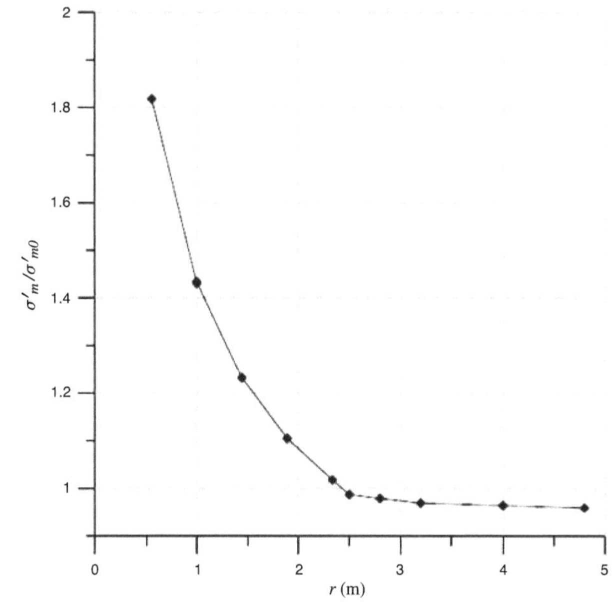

Figure 5.12 Variation of normalized effective mean stress at the end of primary consolidation in soft clay (Mohr Coulomb model) (with permission from the University of Thu Dau Mot, editors Hai et al., 2013)

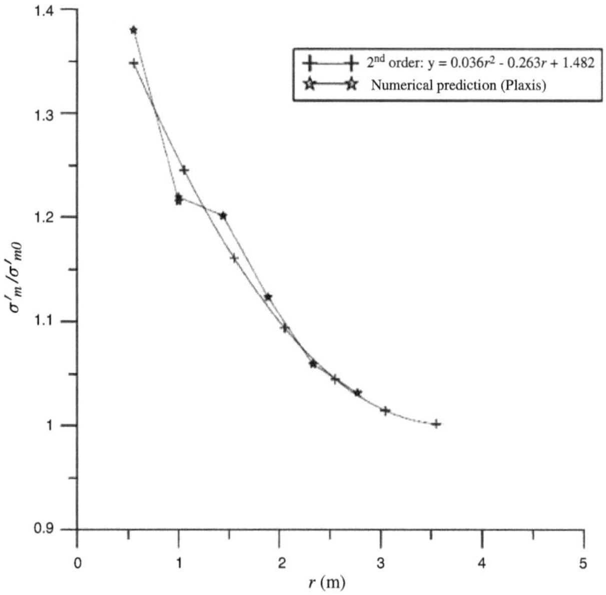

Figure 5.13 Variation of normalized effective mean stress at the end of primary consolidation in soft clay (Hardening Soil Model) (with permission from the University of Thu Dau Mot, editors Hai et al., 2013)

are presented in Appendix 5.1 for normally consolidated soft soil. Numerical simulation by the HSM was performed using data given by Guetif et al. (2008).

Figure 5.13 shows the variation of normalized effective mean stress as a function of the influence zone radius predicted according to the adopted HSM for the soft clay layer.

Figure 5.14 compares the normalized effective mean stresses predicted by the UCM with those obtained by the Mohr Coulomb model and the HSM. Note that the HSM predicts an approximate 12% improvement in Young's modulus of soft clay (averaged value over six times the radius of stone column) while the Mohr Coulomb model predicts as much as a 37.5% improvement. This gap between predictions by Mohr Coulomb and the HSM of improvements in Young's modulus can be explained by the hardening effect. Since this latter is taken into

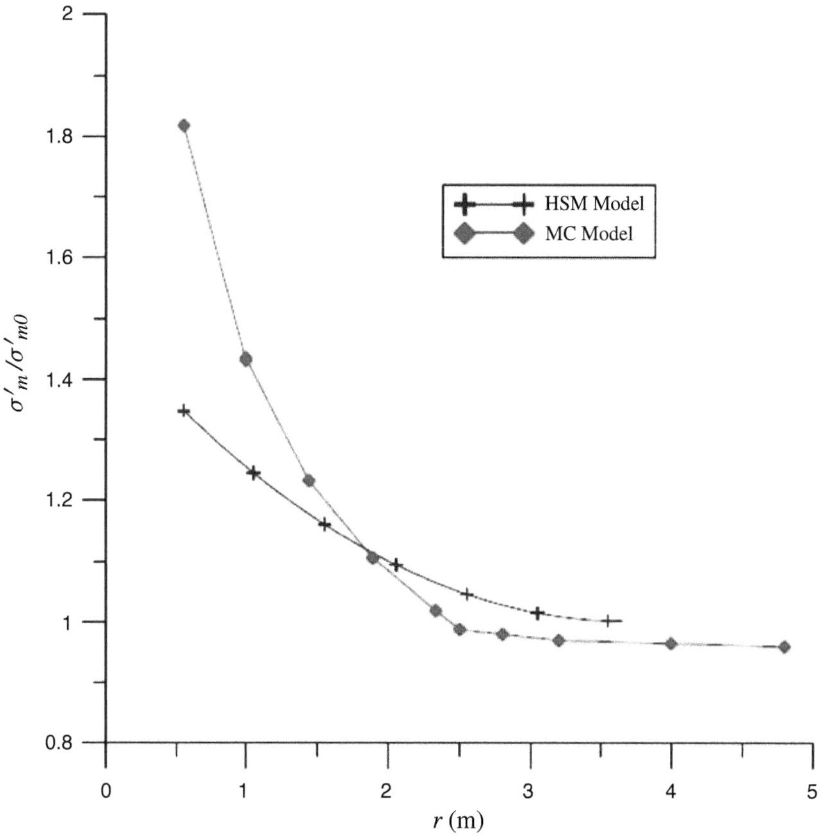

Figure 5.14 Variation of normalized effective stress at mid-thickness of the soft clay layer adopting the Mohr Coulomb and HSM (with permission from the University of Thu Dau Mot, editors Hai et al., 2013)

account by the HSM, the improvement in rigidity will not be as much as that predicted by the Mohr Coulomb model, which does not depend on the hardening effect.

5.3.2 Numerical Simulations Using the Group of Columns Model

Seven stone columns in a triangular pattern were analyzed using the group of columns model (Figure 5.15). Numerical axisymmetric analysis consists of installing a central stone column followed by the installation of a stone crown that is assumed to be equivalent to the six surrounding stone columns. The equivalent thickness of the stone crown, b_{cr}, can be determined according to Equation (5.3):

$$N_c.\pi.a^2 = 2\pi s b_{cr} \tag{5.3}$$

where a = radius of the stone column,

N_c = number of stone columns equivalent to a stone crown (after Figure 5.15: $N_c = 6$)

c = spacing between columns.

The reinforcement effect by the vibro-compacted stone column group is analyzed using the soil profile shown in Figure 5.16. The external radius of the axisymmetric UCM is set equal to 3.05 m. Since the stone columns are installed

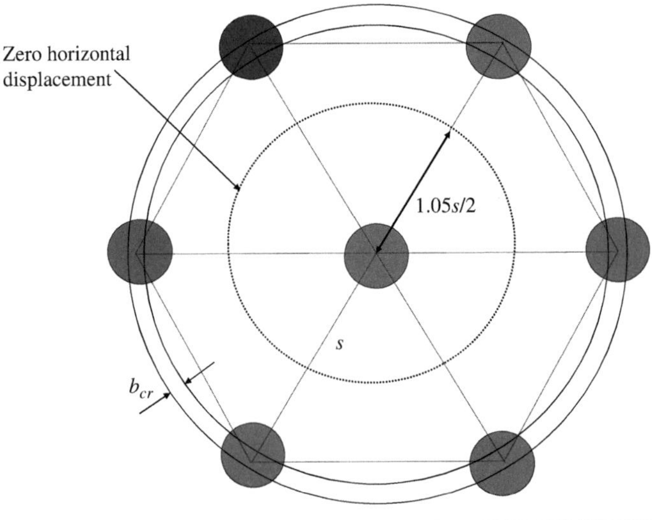

Figure 5.15 Equivalent crown of stone columns (with permission from editor E. Hamdi)

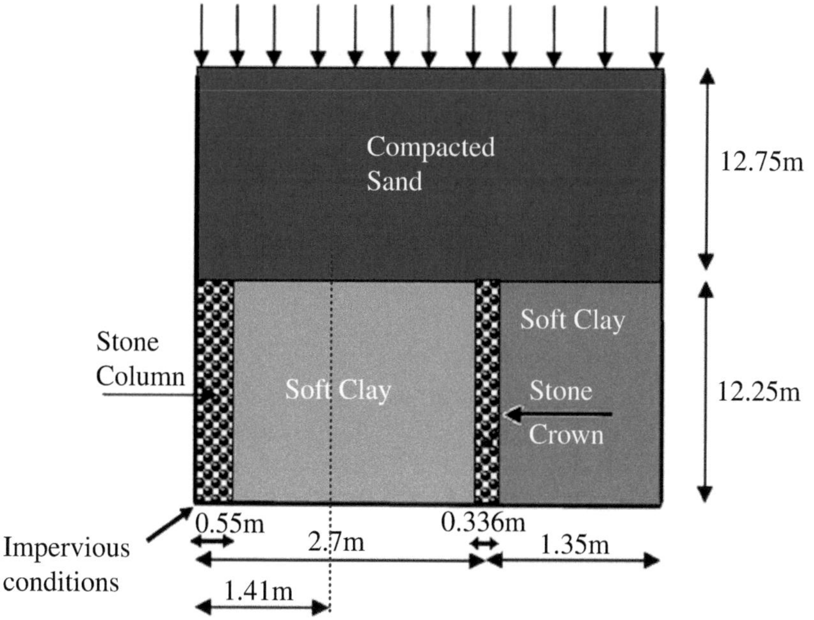

Figure 5.16 Axisymmetric model of a stone column surrounded by a crown of stone (with permission from the University of Thu Dau Mot, editors Hai et al., 2013)

in a very short period of time, the lateral expansion is considered to occur in undrained conditions.

Using the dummy material procedure described in Section 5.3.1, the simulation of stone crown installation corresponds to the horizontal expansion of a cylindrical trench having an initial thickness equal to 0.064 m over the thickness of the soft clay layer. Then, along the circumferences of the simulated trench borders, the soft clay is subjected to a horizontal displacement which simulates the vibro-stone column installation until the horizontal expansion corresponds to an equivalent crown's thickness equal to b_{cr} = 0.336 m. Note that the simulation of installation of the central stone column and the equivalent stone crown takes place in a short-term condition for the execution of the stone columns project.

Once the simulation of lateral expansion has ended, the dummy material is replaced by the stone material (high friction angle and zero cohesion) in the central column and the equivalent crown. Then, the primary consolidation is allowed to occur for a period of 335 days.

When the consolidation of the soft clay due to the group of stone columns installation is over, a uniformly distributed load of 100 kPa is subjected to the reinforced soil model, and then followed by a full primary consolidation for a period of 405 days.

It is important to mention that the model shown in Figures 5.15 and 5.16 has only been considered to predict the improvement in Young's modulus of the soft clay due to the installation of the stone columns group. Such a model is suitable to run the 2D axisymmetric numerical predictions for a group of columns modeled by an equivalent stone crown. Therefore, predictions of the extent of improved zones around the central column and the equivalent crown of stone material will help in optimizing the IAR for the whole design of reinforced soft soil by stone columns. Such a design includes both the bearing capacity and settlement verifications (Bouassida, 2012).

5.3.2.1 Processed Numerical Predictions

Based on the numerical investigation described above, installing a group of columns comprises three successive stages. The first one considers the period of time which immediately follows the horizontal expansion of the columns group. The second stage corresponds to the end of the primary consolidation that takes place in the soft clay as a result of the installation of columns. The third stage marks the end of the consolidation due to the uniformly distributed load applied at the surface of the reinforced soil model.

In order to optimize the quantity of stone material to be incorporated in soft clay, a parametric study has been carried out by varying the spacing between columns. Three spacing values were adopted: $s = 4.9\ a = 2.7$ m; $s = 6\ a = 3.3$ m; and $s = 8\ a = 4.4$ m (a is the radius of the stone column) with respective IARs: $\eta = 0.151$; $\eta = 0.101$; and $\eta = 0.057$.

Results of the parametric study are presented in terms of the extent of influence zone due to columns installation. Further, the settlement of reinforced soil is analyzed when subjected to a uniform load of 100 kPa.

5.3.2.2 Parametric Study for Varied Spacing Between Columns

The distributions of horizontal and vertical displacements are analyzed as well as the change in effective mean stress for varied IAR (or for varied spacing between columns).

Horizontal Displacement

Due to respective lateral expansions that simulate the installation of a central column and the equivalent crown of stone material, horizontal displacement occurs in the surrounding soft clay. The first aim of numerical simulation is to check that no overlap occurs between the influence zones of the central column and that of the crown of stone (e.g., where outward horizontal displacement is predicted). Hence, the evolution of horizontal displacement is analyzed at radius = 1.05 $s/2$, which is equal to 1.41 m, 1.73 m, and 2.3 m, respectively, for the three adopted normalized spacings ($s = 2.7$ m) $s/a = 4.9$, $s/a = 6$, and $s/a = 8$. Those spacings fit within the range of practiced values of the IAR.

Overall, a unique uniform profile of horizontal displacement is predicted immediately after the installation of the stone columns group and at the end of consolidation of improved soft clay.

The predicted horizontal displacement shown in Figures 5.17, 5.18, and 5.19 clearly indicates, dependent less on the depth within the soft clay layer and s/a ratio, that there is no overlap between the influence zones of the installed central column and the surrounding crown of stone.

Vertical Displacement

The installation of seven stone columns generates a significant heave of the ground surface of about 52 cm which corresponds to 4.2% of the total volume of the reinforced soil model shown in Figure 5.16. This can be explained by the adopted elasticity of soft clay, the rapid incorporation with the displacement

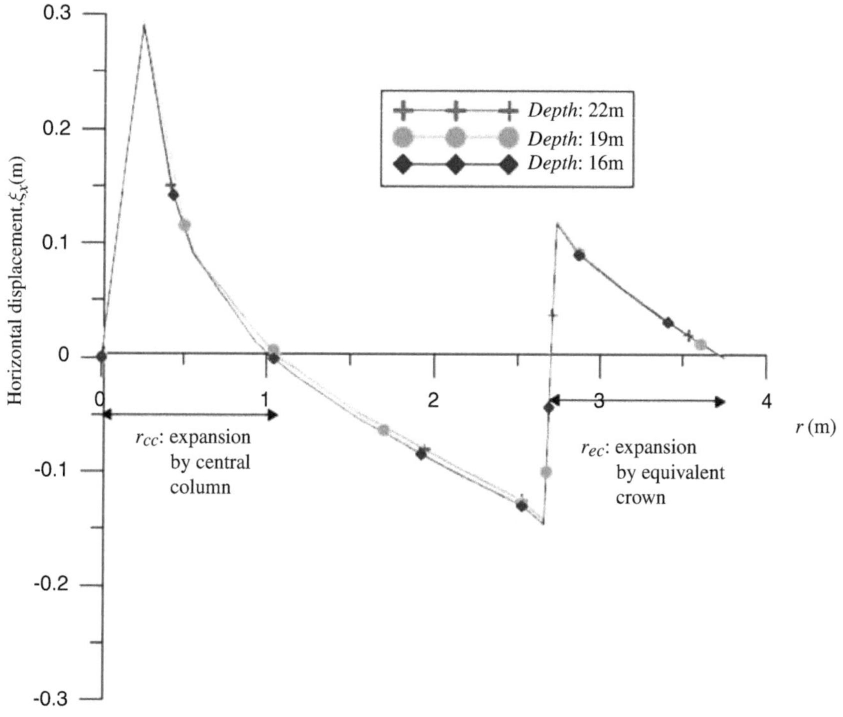

Figure 5.17 Evolution of horizontal displacement at different depths for the case of s = 2.7 m (with permission from the University of Thu Dau Mot, editors Hai et al., 2013)

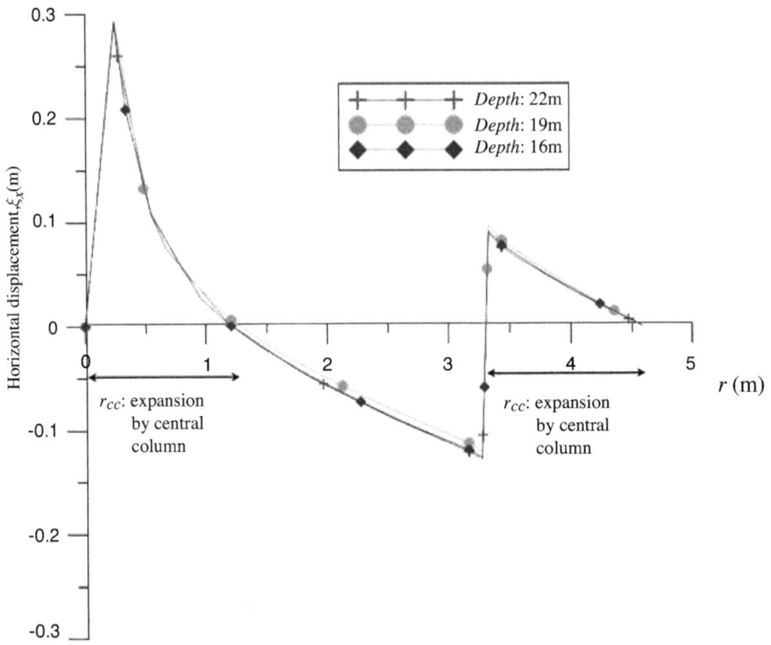

Figure 5.18 Evolution of the horizontal displacement at different depths for the case of s = 3.3 m (with permission from the University of Thu Dau Mot, editors Hai et al., 2013)

method of stone material in quasi-uncompressible soft clay, and the oedometer conditions of the studied UCM (zero horizontal displacement at border).

At the end of the primary consolidation of soft clay, caused by the installation of the stone columns group, a uniform settlement is predicted at the surface of the reinforced soil, while a negligible settlement of the sand layer is estimated.

The reduction in settlement attributed to the installation of the columns group, taking into account the improvement of the soft clay prior to the final loading, is predicted by Plaxis software. When considering a uniform load of 100 kPa, the predicted settlements at the surface of the reinforced soil model are: 15 cm, 17 cm, and 21 cm, for the three adopted normalized spacings: s/a = 4.9; s/a = 6; and s/a = 8, respectively. The settlement of unreinforced soft clay is about 46 cm. Figure 5.20 shows the variation of the settlement at the surface of the reinforced soil model sketched in Figure 5.16, as a function of the IAR. From this figure, one can note that a coefficient of settlement reduction equal to 3 is possible when the improvement ratio is just equal to 15%.

The settlement reduction factor, defined by the ratio between settlements of unimproved soil to that of improved soil (for the same applied load of 100 kPa)

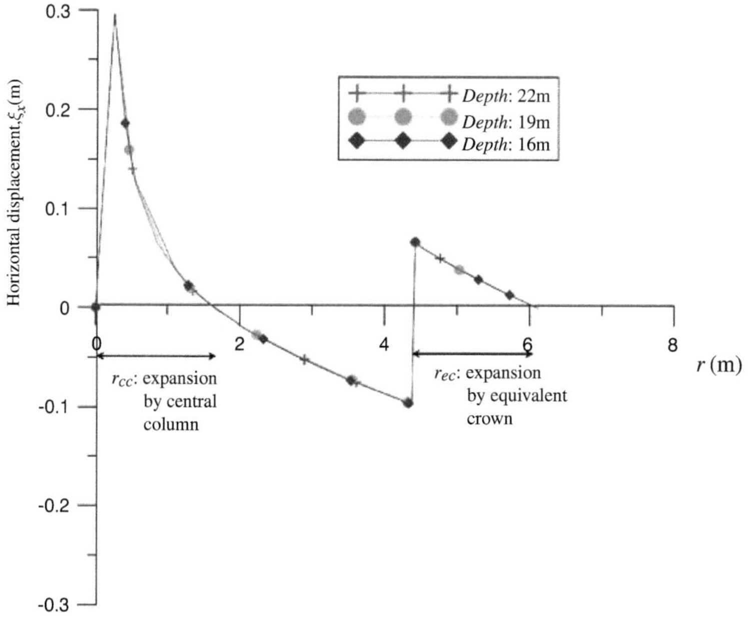

Figure 5.19 Evolution of the horizontal displacement at different depths for the case of s = 4.4 m (with permission from the University of Thu Dau Mot, editors Hai et al., 2013)

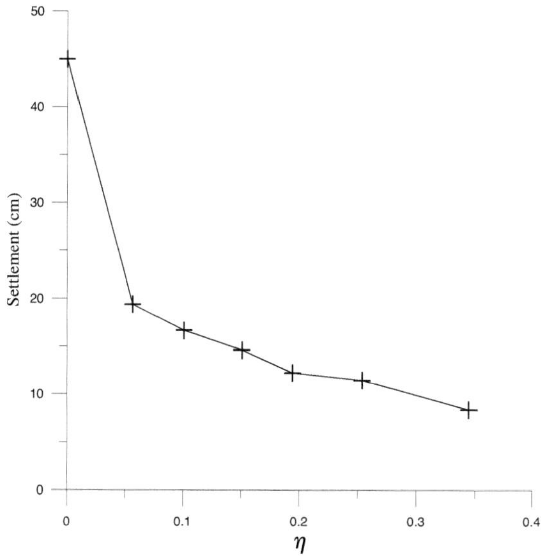

Figure 5.20 Variation of the settlement at the surface of reinforced soil as a function of the IAR (uniform surcharge load is 100 kPa)

is 3.06, 2.7, and 2.19, for the normalized spacing $s/a = 4.9$; $s/a = 6$; and $s/a = 8$, respectively.

Note that for the spacing $s = 2.7$ m, the coefficient of settlement reduction is very effective due to the high confinement, while for the spacing $s = 4.4$ m, it is equal to the settlement reduction factor of the UCM having an external radius of 5 m. Meanwhile, for the spacing $s = 4.4$ m, the settlement reduction has no influence on the group of columns installation.

Variation of Effective Mean Stress

Figure 5.21 shows the variation of effective mean stress, as a function of the horizontal distance, at mid-depth of the soft clay layer for the three adopted

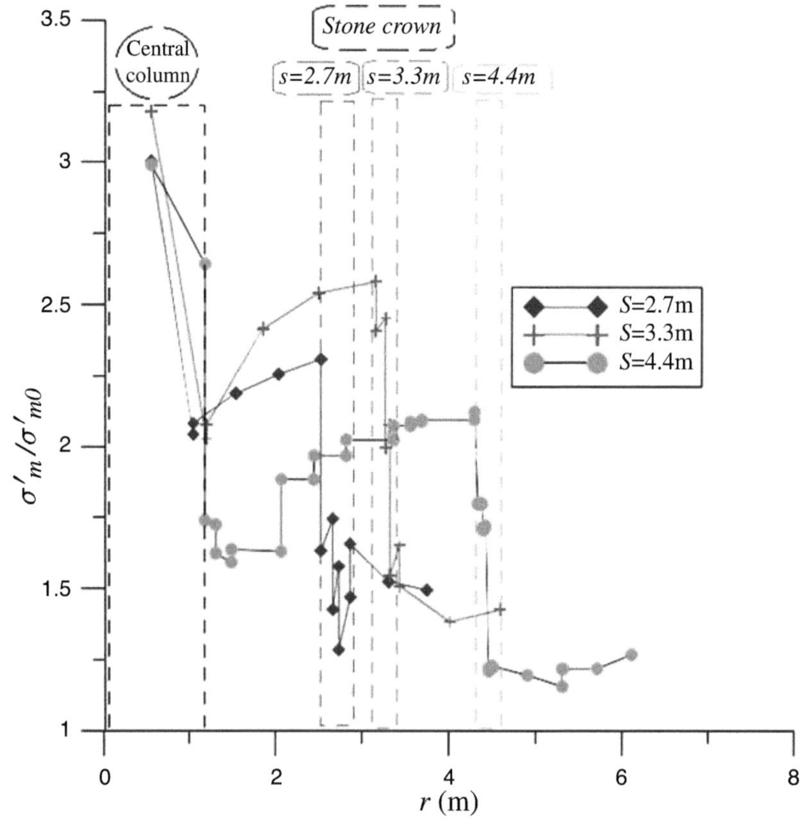

Figure 5.21 Variation of normalized effective mean stresses vs. spacing between stone columns (with permission from the University of Thu Dau Mot, editors Hai et al., 2013)

spacings s =2.7 m; s = 3.3 m; and s = 4.4 m, after a consolidation time of eleven months. Note that a significant increase in effective mean stresses leads to an improved Young's modulus of the soft clay as derived from Equation (5.1).

Table 5.2 presents the increase in mean value of Young's modulus of the soft clay between the central column and the crown stone and beyond the crown stone. The best improvement in Young's modulus is expected between the central column and the stone crown when spacing equals 3.3 m; further, beyond the stone crown the best improvement is predicted for a spacing of 2.7 m.

Predictions of Settlement of Reinforced Soil

The settlement prediction is studied, using numerical computation, at varied IARs (spacing respectively) η = 0.15 (s = 2.7 m); η = 0.10 (s = 3.3 m); and η = 0.057 (s = 4.4 m). Two assumptions were made in order to predict the settlement using each of these models. The first assumption postulates that Young's modulus of the soft clay will be improved due to the installation of stone columns. The second assumption does not consider this improvement of the soft clay properties.

Figure 5.22 compares the settlement variations at the surface of the reinforced soil as predicted by the numerical computations for two cases: first, when the improvement of soft clay properties due to stone column installation is not considered, and second, when such improvement is taken into account. In this latter case, a considerable saving of reinforcing material, for a settlement reduction factor of 3, is expected. Indeed, the IAR decreases from 15% to 5% when the improvements of soft clay properties due to stone column installation are taken into account.

Figure 5.22 also shows that the best settlement reduction is obtained for an IAR of 10%. As a matter of fact, the settlement of reinforced soil decreases from 17 cm to 11 cm when an improved Young's modulus of soft clay is adopted.

Table 5.2 Improvement in Young's modulus of soft clay predicted by the group of stone columns model

Spacing (m)	External radius of equivalent axisymmetric cell (m)	Between central column and stone crown	Beyond stone crown
2.7	3.750	1.86	1.53
3.3	4.589	2.38	1.42
4.4	6.112	1.92	1.22

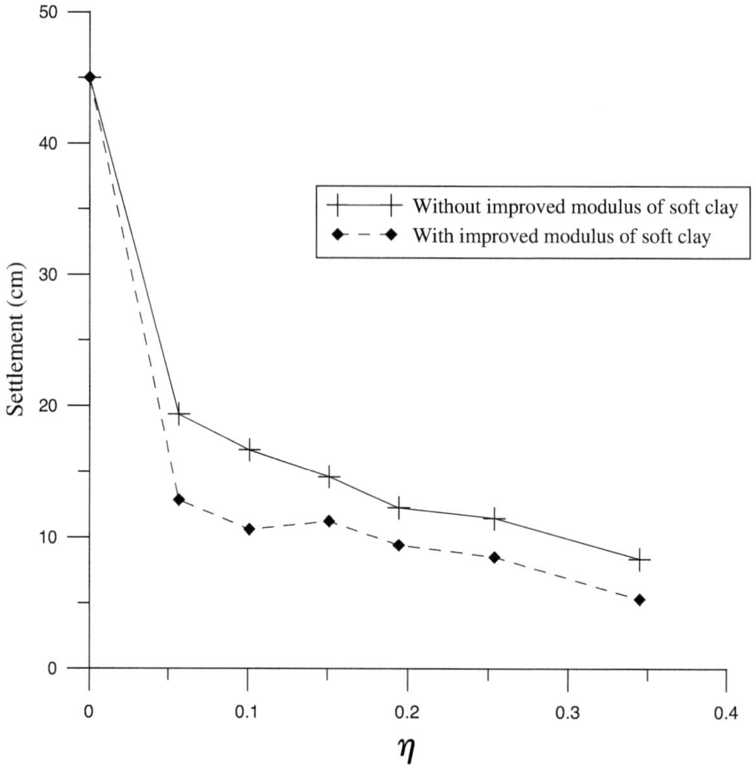

Figure 5.22 Predicted settlements by the group of columns model versus IAR

5.3.3 Experimental Simulation of Stone Column Installation in Soft Clay

The study of experimental behavior of reinforced soft clay by a laterally expanded stone column has been recently investigated by performing triaxial tests on remolded kaolin clay (Bouassida et al., 2012 and Frikha et al., 2013). The proposed experimental procedure simulates the installation of a stone column by means of an applied horizontal pressure within an expanded cylindrical cavity. The tested specimen is a hollow cylinder with an external diameter of 10 cm, an inner diameter of 2 cm, and a height of 20 cm. The triaxial shear test is performed on the specimen to analyze how the mechanical properties of soft clay are modified, depending on the applied expansion pressure.

The tested soft clay was speswhite kaolin; its grain size distribution shows 80% of particle size less than 2 μm (Figure 5.23). Properties of kaolin clay are as follows: specific gravity = 2.6, liquid limit = 57%, plasticity index = 24%, compression index = 0.136, and swelling index = 0.00951 (Frikha et al., 2013).

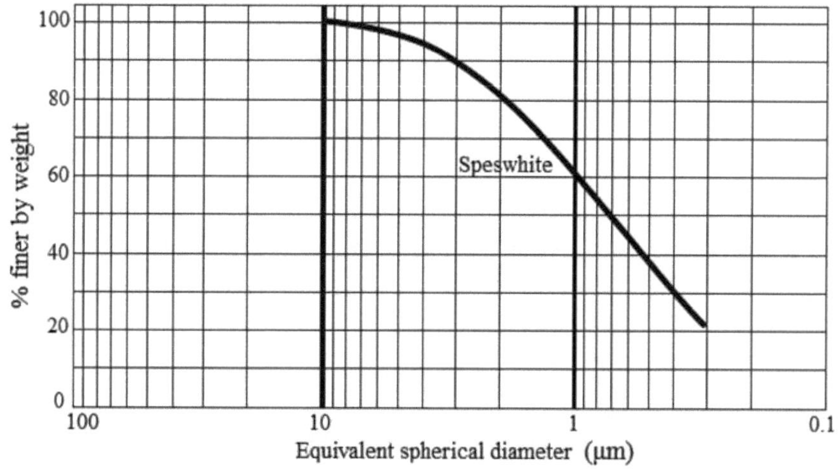

Figure 5.23 Gradation curve of tested kaolin. Geotech. Geol. Eng. J., Observed behaviour of laterally expanded stone column in soft soil, 31 (2), 2013, 739–752, Frikha, W., Bouassida, M., and Canou, J. (With permission from Springer.)

5.3.3.1 Preparation of Clay Specimen

The kaolin specimen is prepared by using the slurry consolidation method. The powder of virgin kaolin is mixed at a water content of 86% that is about 1.5 times its liquid limit. The obtained slurry is shaken by a mechanical mixer for approximately 30 minutes, and then placed in a vacuum bell jar for 24 hours to eliminate any trapped air. The slurry sample is subjected to an initial K_0 preconsolidation using the consolidometer, which is a hollow tube 100 mm in internal diameter and 300 mm in height. A vertical thin tube that is 20 mm in diameter is covered by a filter paper and placed on the axis of the consolidometer to form the cavity. The consolidometer is then filled with the slurry and the sample is formed. The remolded sample is drained vertically by two porous stones placed at the top and bottom sides of the consolidometer. A horizontal drainage is assured by four bands of 40 mm thick filter paper placed at the internal surface of the hollow tube. Silicone grease is spread on the internal contact surface between the remolded soil and the tube to reduce friction effects which allows easy unmolding of the specimen. Using a spoon, the mold is carefully filled with the slurry, which is gently tamped to exclude any penetrated air. Once the predetermined height of the specimen is reached, the consolidation loading cap is placed and the loading starts slowly to provide the desired consolidation stress that is moderately applied (Figure 5.24). The consolidation is assumed completed when the recorded settlement remains quasi-constant. This procedure was adopted for all tests. Two levels of consolidation stress were used: 98 kPa and 170 kPa.

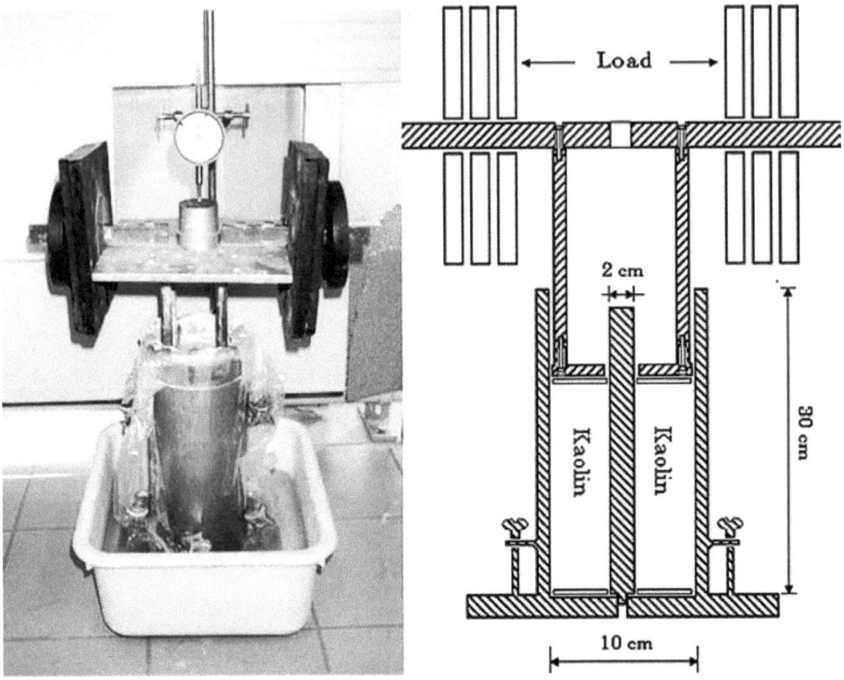

Figure 5.24 K_0 consolidation apparatus. Geotech. Geol. Eng. J., Observed behaviour of laterally expanded stone column in soft soil, 31 (2), 2013, 739–752, Frikha, W., Bouassida, M., and Canou, J. (With permission from Springer.)

5.3.3.2 Expansion of Cylindrical Cavity

The height of the unmolded hollow specimen from the consolidometer varies between 180 and 200 mm. After its installation into the hollow cylindrical cell, the specimen is subjected to a full saturation process, so that the minimum recorded Skempton coefficient is 95%. A detailed description of the hollow cylindrical cell is given in Frikha et al. (2013). The tested isotropic confining stresses were 100, 200, and 300 kPa.

After the consolidation phase, an expansion of the internal cavity is applied under a controlled increase of water volume within it and the corresponding internal pressure by using a pressure volume controller.

The expansion tests are performed in undrained conditions characterized by a prescribed strain rate of 0.055% per minute. According to the in situ installation of stone columns, five levels of cavity expansion are chosen, e.g., $V/V_0 = 1$, 1.25, 1.50, 1.75, and 2.00, where V_0 and V denote the initial- and post-expansion volume of the cavity. It is assumed that the cavity expansion ratios ($V/V_0 = 1.00$,

1.25, 1.50, 1.75, and 2.00) correspond respectively to the durations 0, 7.5, 15, 22.5, and 30 minutes.

During the internal cavity expansion, the applied internal pressure along the cavity border increases. First, this volume expansion was applied to reach an initial volume of cavity V_0, and then when the expansion is pursued, the increase in cavity pressure is recorded while the external confining pressure applied to the specimen remains constant.

As seen in Figure 5.25, the cavity deformation of the specimen may not be perfectly cylindrical; a slight decrease in strain due to the low implementation of the internal pressure is quite difficult, especially when a specimen is pre-consolidated under low stress (less than 50 kPa). In parallel, the final shape of the cavity is also due to the nonrigid type of interior expansion caused by an increase in water volume in a cylinder formed by an elastic membrane. Such a type of loading is slightly affected at the upper and lower bases of hollow cylindrical specimens (border effect).

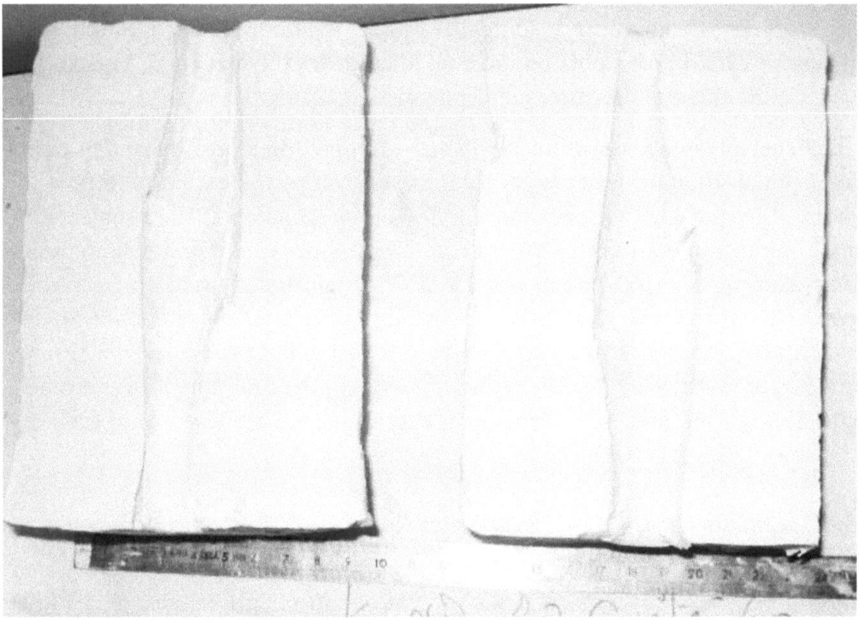

Figure 5.25 Observed failure mode of hollow kaolin specimen subjected to (CU + u) triaxial test. Geotech. Geol. Eng. J., Observed behaviour of laterally expanded stone column in soft soil, 31 (2), 2013, 739–752, Frikha, W., Bouassida, M., and Canou, J. (With permission from Springer.)

5.3.3.3 *Triaxial Tests*

After the expansion of the cavity, the hollow specimen was subjected, in a CU condition with measurement of pore-pressure (CU + u), to an axial displacement rate of 0.03 mm/mn during the shear phase in accordance with the French Standard NF P 94-074 (AFNOR, 1994). Figure 5.25 shows the mode of shear failure produced in a kaolin specimen at the end of the undrained triaxial shear phase.

The experimental program included 15 triaxial tests comprised of the following steps (Frikha et al., 2013):

- Step 1: The specimen is subjected to an isotropic consolidation.
- Step 2: A lateral expansion from the interior of the cavity, in undrained condition, is applied up to a prescribed rate of increase in volume of the specimen (V/V_0 = 1.00, 1.25, 1.50, 1.75, and 2.00) under controlled pressure. The external isotropic stress σ_3 is kept constant. Then, the drainage is allowed with recorded dissipation of excess pore pressure and volume variation of the specimen. This step ends when no volume variation is observed and zero excess pore pressure is reached.
- Step 3: The shear loading is applied, in undrained condition, with recorded excess pore pressure up to failure of the specimen. The internal and external pressures are kept constant.

The behavior of the speswhite kaolin was examined in a post-expanded cavity, for five deformation levels, with total dissipation of the excess pore pressure induced by the cavity expansion. The hollow specimen was then subjected to a CU condition with measurement of the excess of pore pressure at confining pressures equal to 100, 200, and 300 kPa. The majority of tests were performed two or three times to assure the repeatability of test results and, therefore, to validate the experimental set up. Good test repeatability was achieved with a relative error lower than 5%. The reported values of the undrained shear strength and Young's modulus are averages of the calculated results during each test (Frikha et al., 2013).

5.3.3.4 *Influence of Cavity Expansion Ratio*

Figure 5.26 illustrates the variation of the deviatoric stress versus the axial strain for different cavity expansion ratios and under consolidation stress of 200 kPa.

The variation of the peak deviatoric stress as a function of the cavity expansion ratio is shown in Figure 5.27. Note that the peak deviatoric stress increases quasi-linearly with the cavity expansion ratio for different consolidation stress.

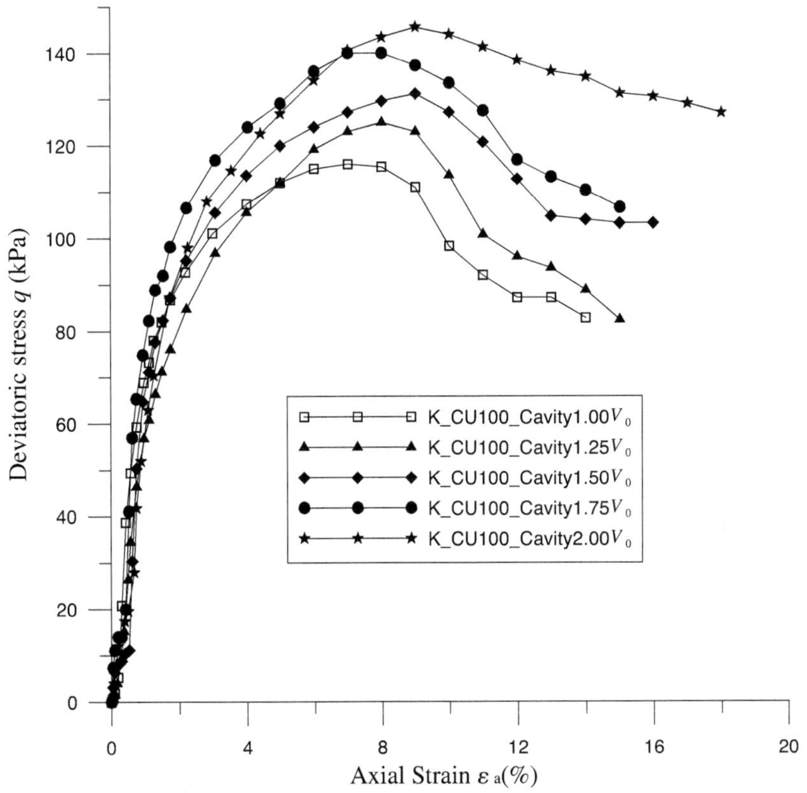

Figure 5.26 Axial stress-strain curves during triaxial shear loading (σ_3 = 200 kPa). Geotech. Geol. Eng. J., Observed behaviour of laterally expanded stone column in soft soil, 31 (2), 2013, 739–752, Frikha, W., Bouassida, M., and Canou, J. (With permission from Springer.)

Figure 5.28 illustrates the variation of normalized peak deviatoric stress q_{max}/σ_3 versus the cavity expansion ratio. This figure clearly indicates that the normalized peak deviatoric stress (q_{max}/σ_3) increases as the cavity expansion ratio V/V_0 increases. Besides, the variation of q_{max}/σ_3 versus the cavity expansion ratio is also more significant at lower consolidation stress. For σ_3 = 100 kPa, q_{max}/σ_3 increases from 0.63 to 0.87 while, for σ_3 = 300 kPa, it increases from 0.53 to 0.62 when the cavity expansion ratio (V/V_0) varies between1.00 and 2.00.

Undrained Shear Strength

From the CU triaxial (CU + u) tests, the undrained shear strength is defined as half of the peak deviatoric stress ($c_u = q_{max}/2$). Figure 5.29 shows the variation

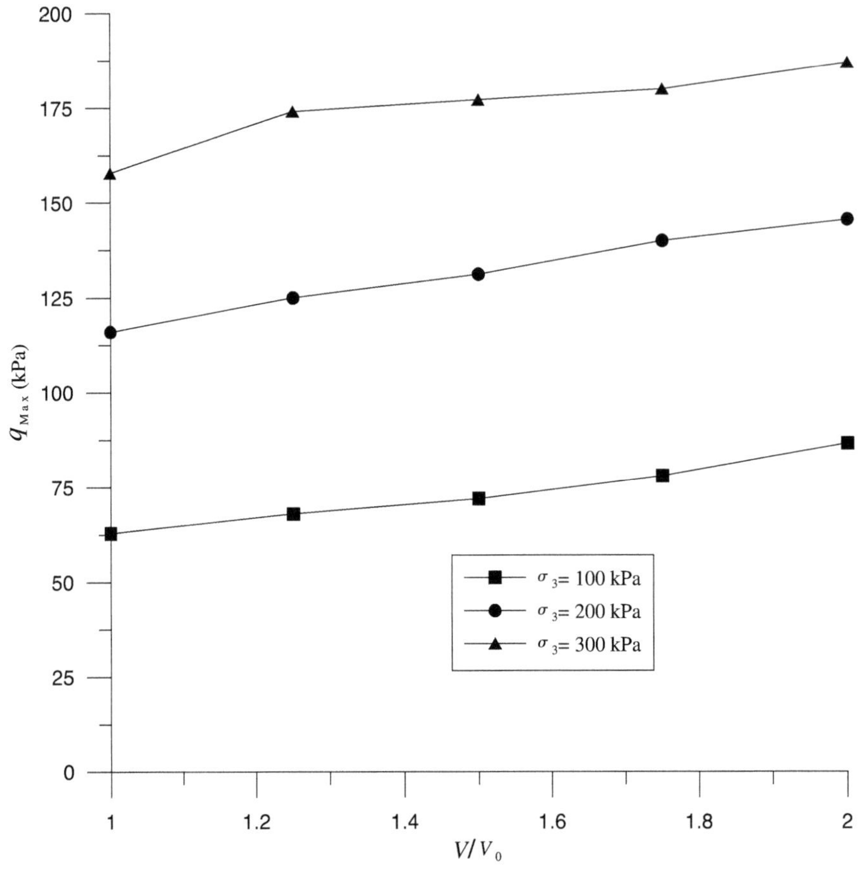

Figure 5.27 Peak values of deviatoric stress versus cavity expansion ratio. Geotech. Geol. Eng. J., Observed behaviour of laterally expanded stone column in soft soil, 31 (2), 2013, 739–752, Frikha, W., Bouassida, M., and Canou, J. (With permission from Springer.)

of the normalized undrained shear strength $c_u(V)/c_u(V_0)$ as a function of the cavity expansion ratio. For the maximum cavity expansion ratio $V = 2V_0$, where V_0 is the initial cavity volume and V the volume of cavity during the expansion process, the undrained shear strength c_u increases by 1.37, 1.25, and 1.18 times for consolidation stresses of 100, 200, and 300 kPa, respectively. The value of the ratio $c_u(V)/c_u(V_0)$ is therefore more significant at lower consolidation stress.

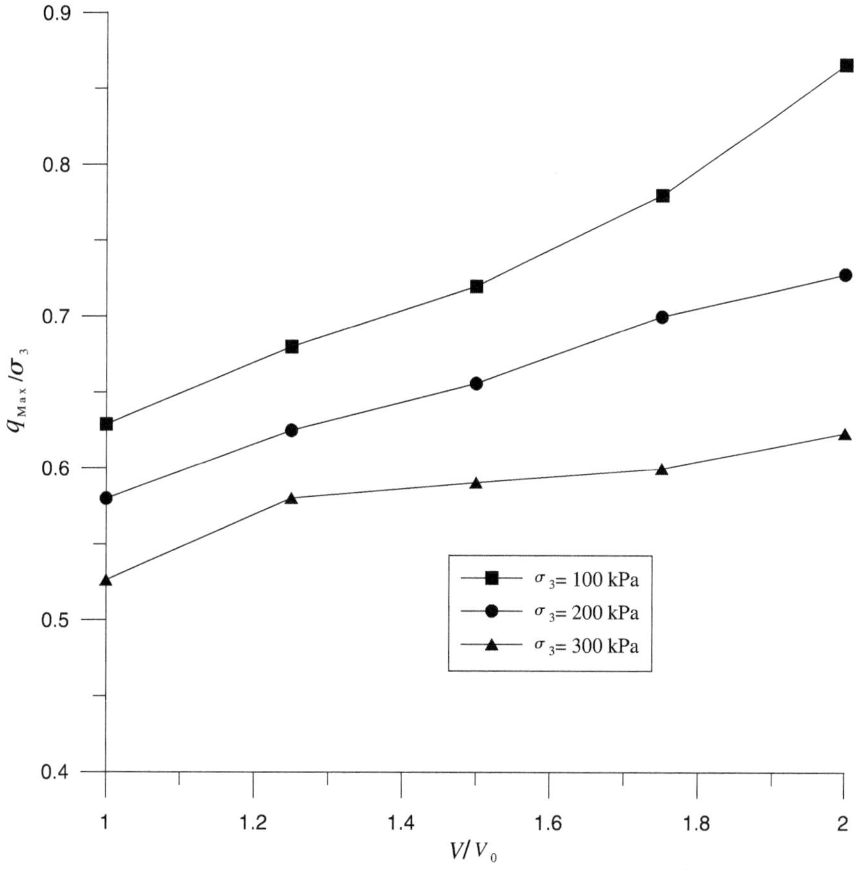

Figure 5.28 Normalized peak deviatoric stress versus cavity expansion ratio. Geotech. Geol. Eng. J., Observed behaviour of laterally expanded stone column in soft soil, 31 (2), 2013, 739–752, Frikha, W., Bouassida, M., and Canou, J. (With permission from Springer.)

Undrained Young's Modulus

The values of undrained Young's modulus can be deduced from the initial quasi-linear behavior of soil in the range of small strains. The shear load phase is performed in a controlled strain rate mode.

The undrained secant Young's modulus is defined as $E = \Delta\sigma/\Delta\varepsilon$ at an axial strain value of 0.01%. If the deviatoric stress-axial strain curve presents, at the beginning, a change in slope due especially to the installation of the specimen and seating errors, Young's modulus is measured immediately after this first phase at the same strain value.

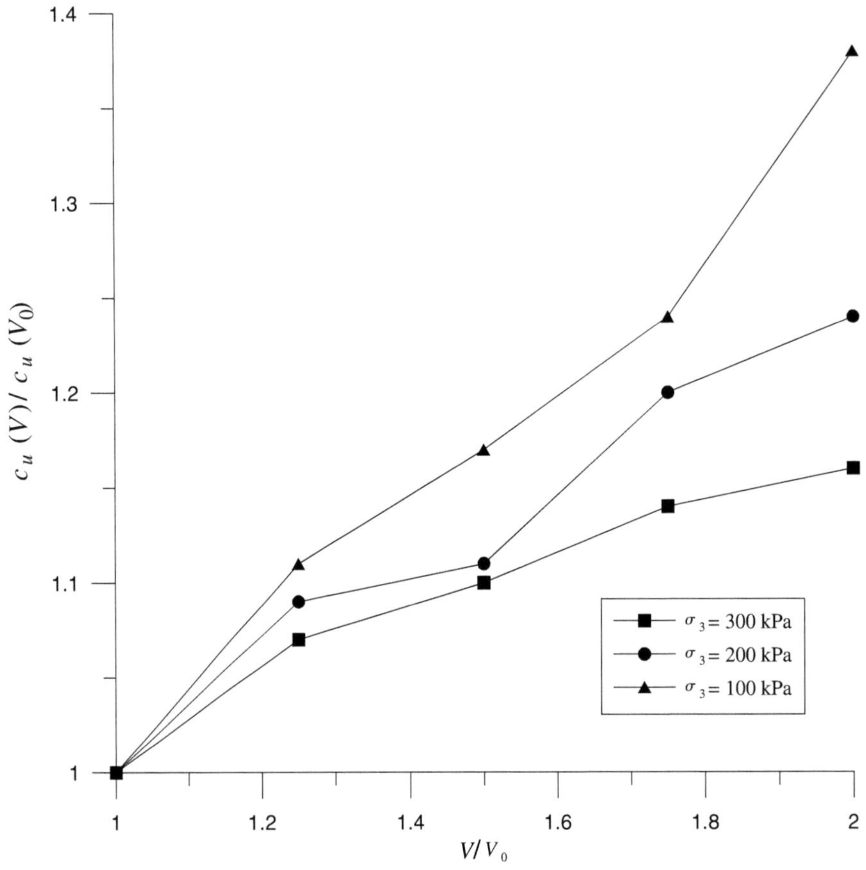

Figure 5.29 Gain in undrained shear strength versus cavity expansion ratio. Geotech. Geol. Eng. J., Observed behaviour of laterally expanded stone column in soft soil, 31 (2), 2013, 739–752, Frikha, W., Bouassida, M., and Canou, J. (With permission from Springer.)

As shown in Figure 5.30, the increase in Young's modulus is clearly noted when both the cavity expansion ratio and the consolidation stress are high. Also, it can be observed that for significant consolidation stress, σ_3 = 200 kPa and 300 kPa, the change of normalized undrained Young's modulus E_u $(V)/E_u$ (V_0) remains quasi-constant when the cavity expansion ratio is greater than 1.5. As depicted in Figure 5.31, at a maximum cavity expansion ratio ($V/V_0 = 2$), the undrained Young's Modulus E_u increased by 1.74, 1.55, and 1.52 times for the consolidation stresses of 100, 200, and 300 kPa, respectively.

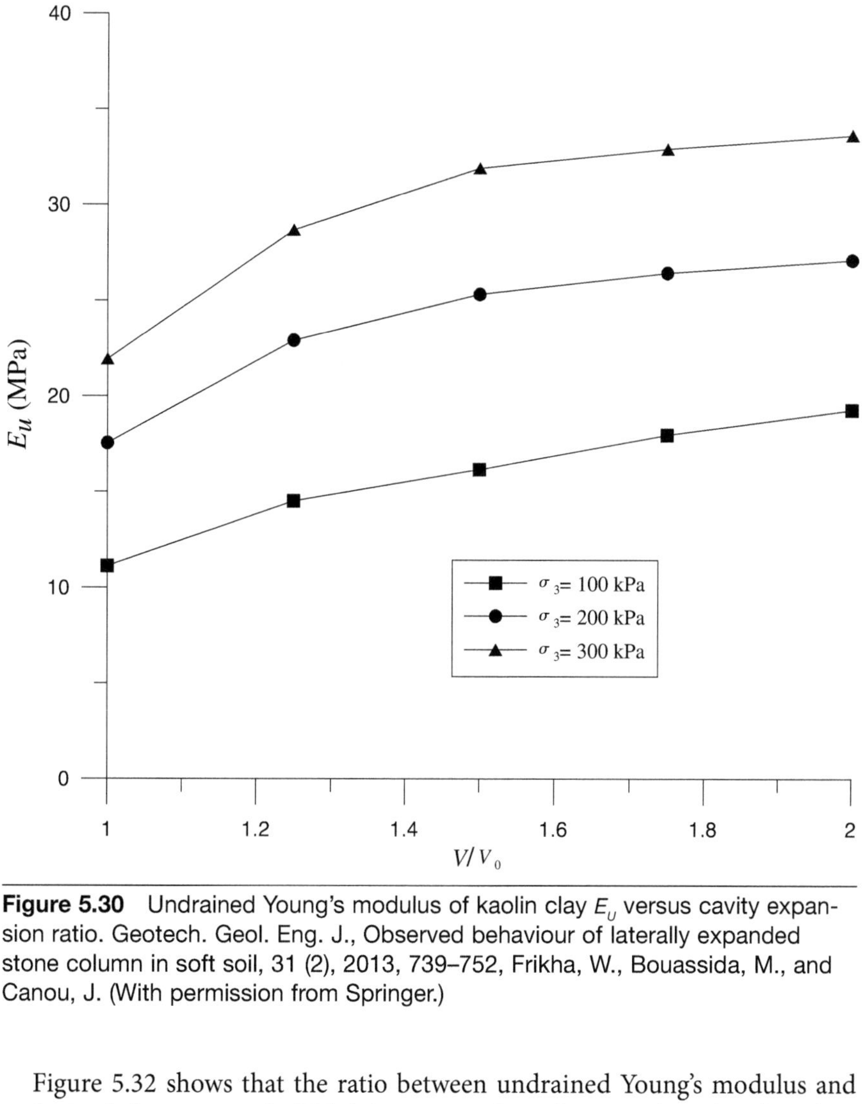

Figure 5.30 Undrained Young's modulus of kaolin clay E_U versus cavity expansion ratio. Geotech. Geol. Eng. J., Observed behaviour of laterally expanded stone column in soft soil, 31 (2), 2013, 739–752, Frikha, W., Bouassida, M., and Canou, J. (With permission from Springer.)

Figure 5.32 shows that the ratio between undrained Young's modulus and undrained shear strength (E_u/c_u) increases when the ratio (V/V_0) is less than 1.50. Note that the higher the ratio (E_u/c_u), the lower the consolidation stress.

Drained Friction Angle

Triaxial test results are currently interpreted from stress paths drawn either in terms of total stress plane (P, q) or in terms of effective stress (P', q). P is the reduced mean total (respectively effective) stress as $P = \frac{\sigma_1 + \sigma_3}{2}$ ($P' = P - u$, "u"

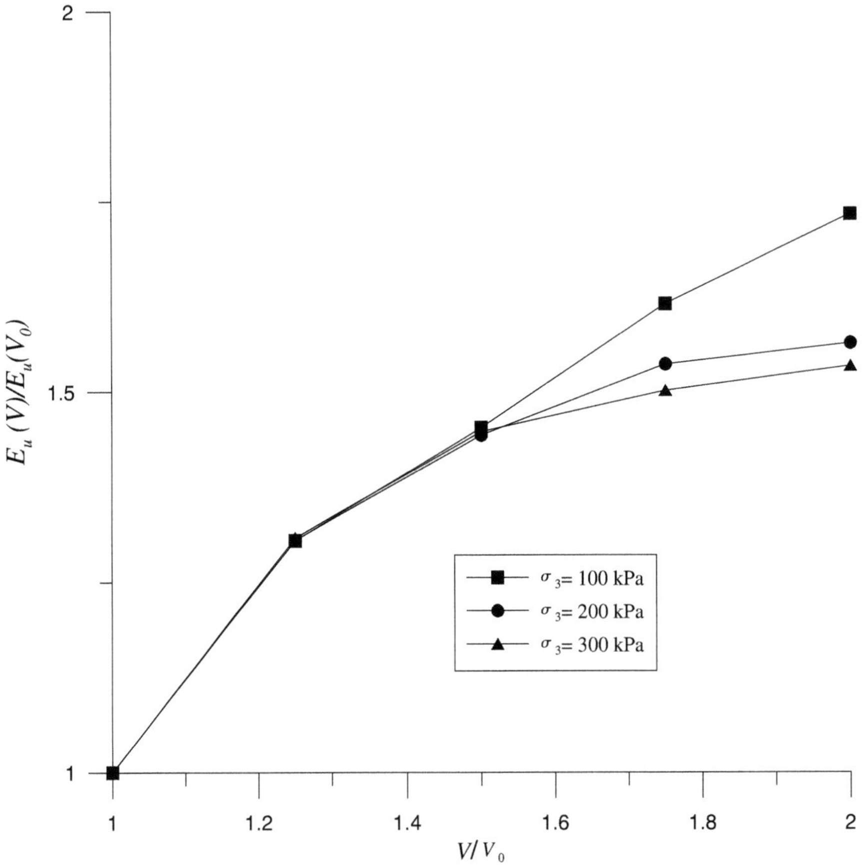

Figure 5.31 Normalized undrained Young modulus versus cavity expansion ratio. Geotech. Geol. Eng. J., Observed behaviour of laterally expanded stone column in soft soil, 31 (2), 2013, 739–752, Frikha, W., Bouassida, M., and Canou, J. (With permission from Springer.)

is the recorded excess pore pressure) and q is the deviatoric stress. Figures 5.33a and b show the measured total and effective stress paths during the undrained shear phases for cavity expansion ratios of $V/V_0 = 1.00$ and $V/V_0 = 2.00$, during which the failure of the tested soil sample occurs at constant volume. The corresponding critical state of stresses defines the critical state characterized by the drained angle of internal friction of the tested soil. The slope of the critical state line shown in Figures 5.33a and b leads to the determination of the drained angle of internal friction.

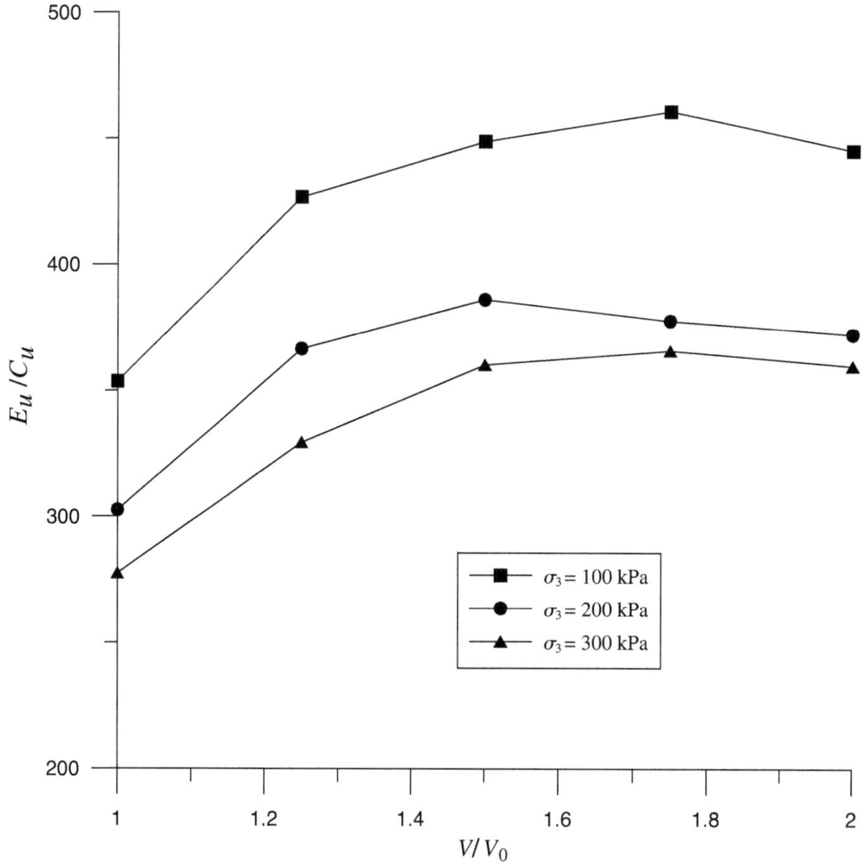

Figure 5.32 E_u/c_u versus cavity expansion ratio. Geotech. Geol. Eng. J., Observed behaviour of laterally expanded stone column in soft soil, 31 (2), 2013, 739–752, Frikha, W., Bouassida, M., and Canou, J. (With permission from Springer.)

Figure 5.34 shows an increase in φ' from 22.1° to 23.6° when the cavity expansion ratio (V/V_0) increases from 1.0 to 2.0. In addition, the increase in φ' is much more significant when the cavity expansion ratio (V/V_0) is lower than 1.50. For instance, when (V/V_0) increases from 1.50 to 2.00, the increase in φ' is negligible.

Based on the measured data, the drained angle of internal friction φ' (in degrees) of remolded kaolin could be expressed as a function of the ratio (V/V_0) as given by Equation (5.4).

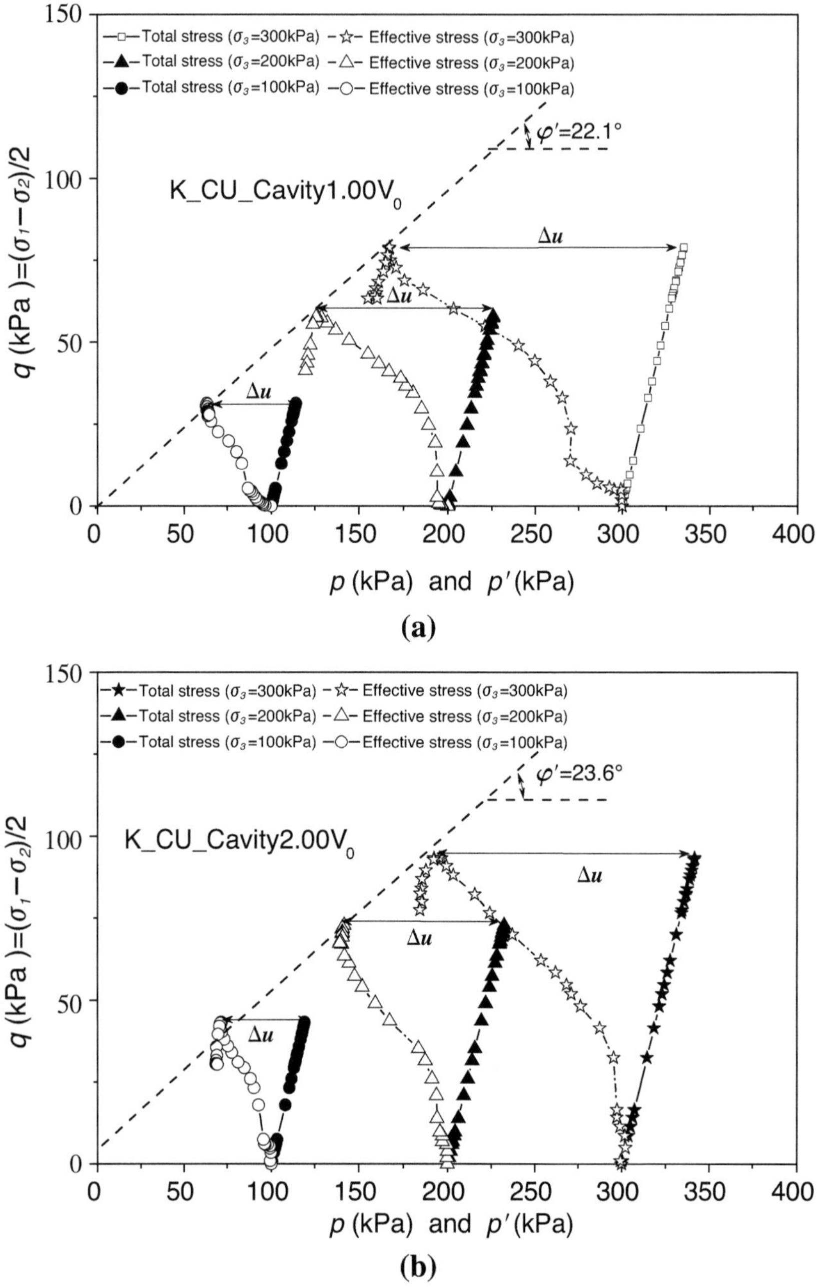

Figure 5.33 Recorded total and effective stress paths during CU + u triaxial tests. Geotech. Geol. Eng. J., Observed behaviour of laterally expanded stone column in soft soil, 31 (2), 2013, 739–752, Frikha, W., Bouassida, M., and Canou, J. (With permission from Springer.)

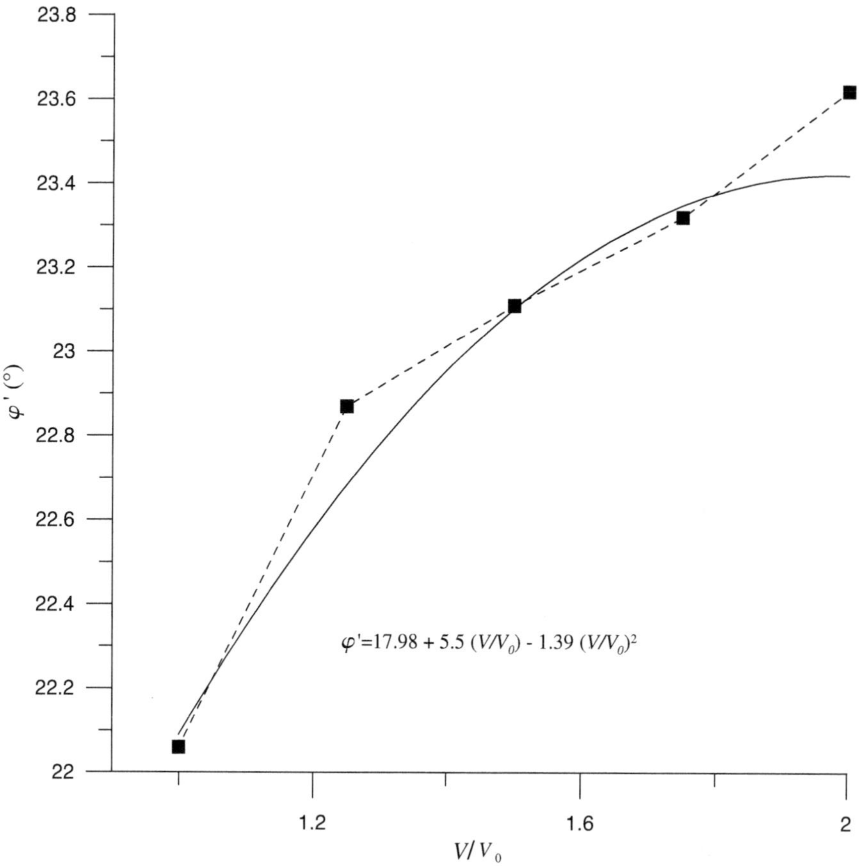

Figure 5.34 Variation of friction angle versus cavity expansion ratio. Geotech. Geol. Eng. J., Observed behaviour of laterally expanded stone column in soft soil, 31 (2), 2013, 739–752, Frikha, W., Bouassida, M., and Canou, J. (With permission from Springer.)

$$\varphi' = -1.39\left(\frac{V}{V_0}\right)^2 + 5.56\left(\frac{V}{V_0}\right) + 17.98 \tag{5.4}$$

It is then concluded, from the experimental study described above, that short-term as well as long-term stiffness and strength characteristics of soft clay are increased in post-lateral expansion that simulates the stone column installation in the field. Increased stiffness and strength characteristics essentially depend on the initial consolidation stress and the cavity expansion ratio as illustrated by Figures 5.31, 5.32, and 5.34. The cavity expansion ratio describes the intensity of stone material installation that influences the extent of improved soft clay around the column.

It should be noted here that when the ratio (V/V_0) is less than 1.50 (the most expected case in practice), the increase in undrained Young's modulus does not depend on the value of consolidation stress. Referring to the numerical predictions primarily proposed by Guetif et al. (2007) and presented in Figure 5.12, it is expected that an average improvement of Young's modulus by 32% will occur around the stone column in a horizontal distance of three to four times its radius. Adopting an estimated normalized undrained Young's modulus $\frac{E_u(V)}{E_u(V_0)}=1.30$, from Figure 5.31, the rate of expanded volume in soft clay is equal to 1.25. This result, to some extent, shows the usefulness of both the numerical and experimental simulations in estimating quite comparable predictions of the soft soil improvement due to stone column installation. Nevertheless, such comparison needs more assessment, for example, by testing different soft soils.

5.4 ACCELERATION OF SETTLEMENT IN POST-FINAL LOADING

Guetif et al. (2004) and Guetif-Fessi and Bouassida (2005) proposed an analytical method for the prediction of settlement and excess pore pressure occurring in improved soft clay by stone columns when subjected to a prescribed history of loading. For this purpose, using a UCM (as sketched in Figure 3.3), a poroelastic model was considered for fully saturated soft clay with a perfectly drained column material.

5.4.1 Equations of Linear Poroelastic Behavior

Equations governing the linear poroelastic behavior of saturated soils with assumed noncompressible constituents (skeleton and water) are briefly presented (Coussy, 1991; Dormieux and Bourgeois, 2002).

The first state equation, Equation (5.5), expresses that the volume variation of the saturated soil $tr(\underline{\underline{\varepsilon}})$ is only due to the water exchange between the soil itself and the outside:

$$tr(\underline{\underline{\varepsilon}})=\frac{m}{\rho_w} \tag{5.5}$$

where m is the supply of water per unit volume of soil, and ρ_w is the specific mass of water.

Consider the isotropic linear elastic behavior with initial natural state ($\underline{\underline{\sigma}}=0$), the state of strains $\underline{\underline{\varepsilon}}$ is governed by the classical Terzaghi effective state of stresses $\underline{\underline{\sigma}}'$ equation:

$$\underline{\underline{\sigma}}' = \lambda' tr\left(\underline{\underline{\varepsilon}}\right)\underline{\underline{I}} + 2\mu'\underline{\underline{\varepsilon}} \tag{5.6}$$

where λ' and μ' are the coefficients of Lamé of the soil skeleton expressed as a function of the drained Young's modulus and Poisson's ratio as given by Equations (3.5a) and (3.5b), where $\underline{\underline{I}}$ denotes the identity tensor.

According to Darcy's law, a linear relationship is established between the vector of fluid mass transfer $\underline{w}$ and the gradient of the excess pore water pressure u:

$$\frac{\underline{w}}{\rho_w} = -k_h grad\left(\frac{u}{\rho_w g}\right) \tag{5.7}$$

where g is the acceleration of gravity and k_h is the soil horizontal coefficient of permeability.

A water mass balance equation expresses the relationship between the rate of water unit supply, $\dot{m}$, and the vector of fluid mass transfer $\underline{w}$:

$$\dot{m} + div\underline{w} = 0 \tag{5.8}$$

When substituting Equations (5.5) and (5.7) into Equation (5.8), one obtains the diffusion equation:

$$\rho_w g tr\left(\dot{\underline{\underline{\varepsilon}}}\right) = k_h \Delta u \tag{5.9}$$

where Δ denotes the Laplace operator.

5.4.2 Statement of the Auxiliary Problem

Due to the symmetry of column geometry, cylindrical coordinates (r, z) are considered in order to define the mechanical and hydraulic boundary conditions of the UCM (Figure 3.3). Mechanical boundary conditions can be written as follows.

The bottom cross section of the UCM (z = 0) is in smooth contact with a rigid and fixed plate:

$$T_r = T_\theta = 0 \qquad \xi_z = 0 \tag{5.10}$$

The top cross section ($z = H$) is in smooth contact with a rigid plate subjected to a downward uniform vertical displacement δ as a function of time t:

$$T_r = T_\theta = 0 \qquad \xi_z = -\delta(t) \tag{5.11}$$

The lateral surface ($r = b$) is in smooth contact with a fixed border.

$$T_z = T_\theta = 0 \qquad \xi_r = 0 \tag{5.12}$$

Hydraulic conditions take into account the high permeability of column material with respect to that of the surrounding soft clay:

$$u(0 \le r \le a, 0 \le z < H) = 0 \tag{5.13a}$$

The excess pore-water pressure at the top horizontal boundary of the soft clay ($z = H$) is zero:

$$u(0 \le r \le a, z = H) = 0 \tag{5.13b}$$

The top and bottom horizontal boundaries of the soft clay ($z = 0, H$), as well as the lateral surface ($r = b$) are impervious:

$$\underline{w}(a \le r \le b, z = H).\underline{e}_z = 0 \tag{5.14a}$$

$$\underline{w}(a \le r \le b, z = 0).\underline{e}_z = 0 \tag{5.14b}$$

$$\underline{w}(r = b).\underline{e}_r = 0 \tag{5.14c}$$

5.4.3 Solution of the Poroelastic Problem

It is assumed that water flow occurs only in a horizontal direction:

$$\underline{w} = w(r,t)\ \underline{e}_r \tag{5.15}$$

For the sake of simplicity, one can introduce the *equivalent membrane* concept, considering that the permeability of soft clay is infinite. As such, the excess pore-water pressure remains homogenous within the soft clay surrounding the column (Figure 5.35), then:

$$a < r \le b, \quad u(r,t) = u_s(t) \tag{5.16}$$

As a result of the assumption in Equation (5.16), the horizontal water flow toward the column will be controlled by the equivalent membrane having the permeability K with unit of invert of time (s^{-1}). Then the water flow is expressed by:

$$\frac{w}{\rho_w}(a,t) = -K\frac{u_s(t)}{\rho_w g} \tag{5.17}$$

Following the above conditions, the solution of the poroelastic problem reduces to that of an elastic problem related to the UCM subjected to a uniform displacement $\delta(t)$ at the top surface and an isotropic state of stress, corresponding to the excess pore pressure in the soft clay. This leads to the following differential equation (Fessi-Guetif, 2004):

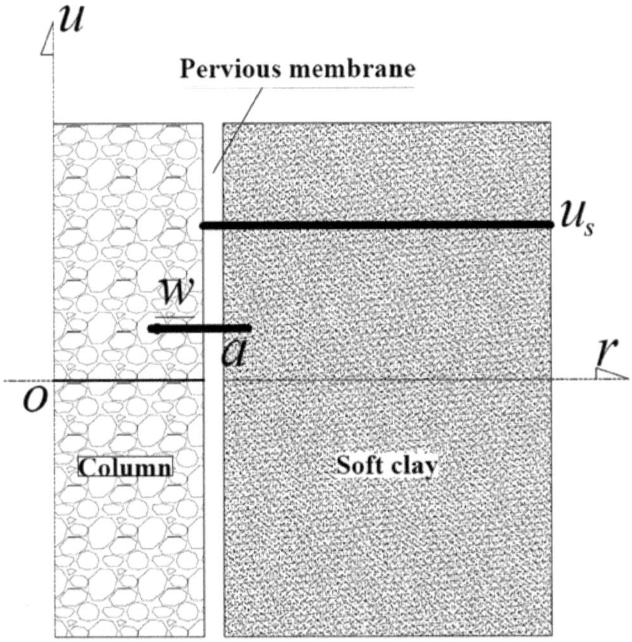

Figure 5.35 Equivalent membrane at the column-soft clay interface. (Reprinted from Proc. 16th Int. Conf. of Soil Mech. and Geot. Eng. September 12th–16th, Osaka, Japan, Vol. 3, Guetif Fessi, Z. & Bouassida, M. Analytical estimate of settlement evolution of soft soil reinforced by stone columns, 1355–1358, Copyright 2005, with permission from IOS Press.)

$$abD_u(a)\frac{\dot{u}_s}{E} - K\frac{a}{\rho_w g}u_s = \left[\frac{a^2-b^2}{2} - abD_d(a)\right]\frac{\dot{\delta}}{H} \tag{5.18}$$

The expressions $D_u(a)$ and $D_d(a)$ are detailed in Appendix 5.2.

From the differential Equation (5.18) and for a given function $\delta(t)$, the development of excess pore-water pressure within the soft clay can be calculated. For the particular case when the top of the UCM is subjected to a uniform settlement of the form $\delta(t)=\delta_0 h(t)$, where $h(t)$ is the Heaviside function, the excess pore pressure can be written as:

$$u_s(t)=\left(\frac{Z}{X}\frac{\delta_0}{H}E\right)\exp\left[-\frac{t}{T_c}\right]h(t) \tag{5.19}$$

where T_c denotes the characteristic time of consolidation that is defined by:

$$T_c = -\frac{\rho_w g b D_u(a)}{KE} \tag{5.20}$$

where E denotes the soft clay Young's modulus and expressions of functions Z and X are given in Appendix 5.2.

The coefficient of permeability K of the equivalent membrane is introduced by $K=\frac{k_h}{L_k}$, where L_k is a characteristic length which is determined from the solution of the horizontal consolidation problem of the UCM. For such a problem (Figure 5.35), the characteristic length will be derived by identifying the characteristic time predicted from Equation (5.20) to that proposed by Barron's theory (1947). The characteristic length is written as:

$$L_k = \frac{k_h bE}{c_h D_u(a)\rho_w g}\left[\frac{\ln\left(\sqrt{\eta}\right)}{2(1-\eta)}+\frac{3-\eta}{8}\right] \quad (5.21)$$

c_h = the coefficient of horizontal consolidation as introduced by Barron (1947). The IAR is given by $\eta=a^2/b^2$ as previously introduced by Equation (1.1).

5.4.4 Settlement Prediction for a Given Loading History

In practice the main problem is to solve the prediction of the settlement evolution generated by a given loading history. The load *Q(t)*, exerted by the assumed rigid rafted foundation on the reinforced soil is:

$$Q(t)=-\int_{S(z=H)}\sigma_z(H,t)dA \quad (5.22)$$

σ_z is the vertical stress distribution at the top of the UCM. Calculation of this vertical load (detailed in Fessi-Guetif, 2004) leads to:

$$Q(t)=\pi b^2\left\{\begin{array}{l}\left[\left(\eta(\lambda_c+2\mu_c)+(1-\eta)(\lambda+2\mu)\right)-2(\lambda_c-\lambda)\frac{a}{b}D_d(a)\right]\frac{\delta(t)}{H}\\+\left[(1-\eta)E-2(\lambda_c-\lambda)\frac{a}{b}D_u(a)\right]\frac{u_s(t)}{E}\end{array}\right\} \quad (5.23)$$

Combining Equations (5.19) and (5.23), Equation 5.24 could be formulated as:

$$\theta\,\dot{Q}+\chi\,Q=\omega\dot{\Delta}+\psi\left(\frac{\delta(t)}{H}\right) \quad (5.24)$$

The expressions of coefficients θ, χ, ω and ψ are detailed in Appendix 5.2. Given the loading history *Q(t)*, Equation (5.24) can be solved by using the Laplace-Carson's transform in the same way as for solving a viscoelastic problem (Salençon, 1983).

In the particular case of an instantaneous loading for $t = 0$:

$$Q(t)=Q_0\quad h(t) \quad (5.25)$$

The solution in terms of normalized settlement could be written as:

$$\left(\frac{\delta}{H}\right)(t)=Q_0\left[\frac{\theta}{\omega}\exp\left(-\frac{\psi}{\omega}t\right)+\frac{\chi}{\psi}\left(1-\exp\left(-\frac{\psi}{\omega}t\right)\right)\right]h(t) \qquad (5.26)$$

The development of settlement and the degree of horizontal consolidation for a prescribed load subjected to the foundation may be calculated as functions of the IAR and the linear elastic characteristics of the UCM's constituents.

By the use of the poroelastic model proposed by Equation (5.26), the settlement variation versus time of the reinforced soil due to the instantaneous loading can be predicted. This result is illustrated in Figure 5.36 as a function of the IAR.

Furthermore, the excess pore pressure development $u_s(t)$ and the horizontal degree of consolidation development U_h (%), as defined in Fessi-Guetif (2004), are also predicted (Figure 5.37).

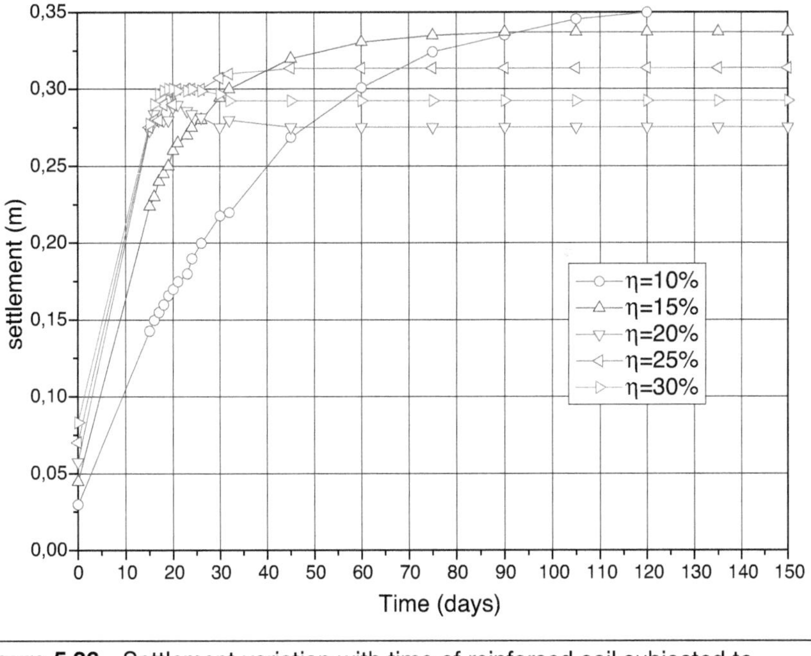

Figure 5.36 Settlement variation with time of reinforced soil subjected to instantaneous loading (E =10^3 kPa, E_c = 10^4 kPa, Q = 10^3 kN, k = 10^{-8} m/s, b = 1.7 m, H = 15m). (Reprinted from Proc. 16th Int. Conf. of Soil Mech. and Geot. Eng. September 12th–16th, Osaka, Japan, (3), Guetif Fessi, Z. & Bouassida, M. Analytical estimate of settlement evolution of soft soil reinforced by stone columns, 1355–1358, copyright 2005, with permission from IOS Press.)

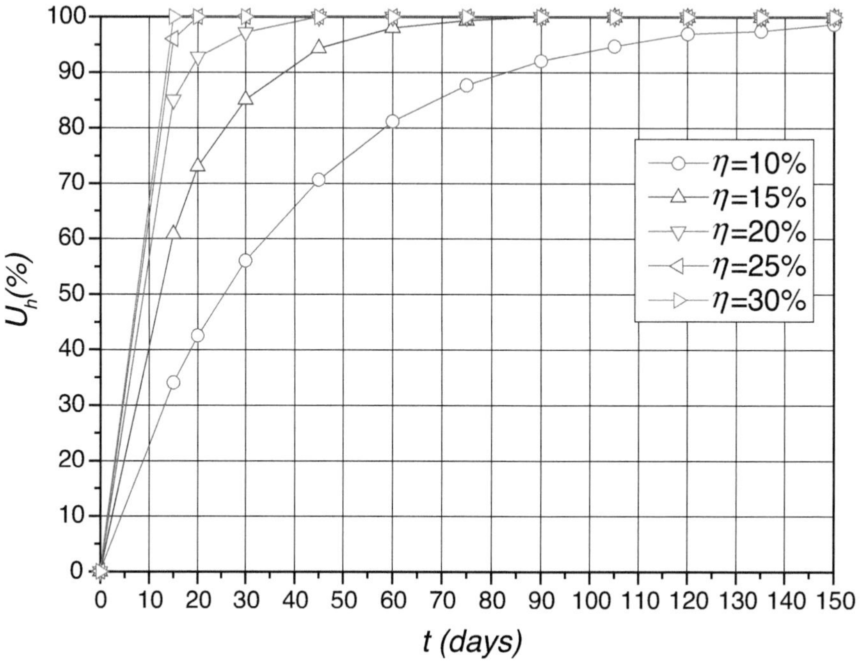

Figure 5.37 Variation of horizontal consolidation degree vs. time of reinforced soil subjected to instantaneous loading ($E = 10^3$ kPa, $E_c = 10^4$ kPa, $Q = 10^3$ kN, $k_h = 10^{-8}$ m/s, $b = 1.7$ m, $H = 15$ m. (Reprinted from Proc. 16th Int. Conf. of Soil Mech. and Geot. Eng. September 12th–16th, Osaka, Japan, (3), Guetif Fessi, Z. & Bouassida, M. Analytical estimate of settlement evolution of soft soil reinforced by stone columns, 1355–1358, copyright 2005, with permission from IOS Press.)

Figure 5.38 illustrates the prediction made by the poroelastic approach with measured settlement for the case history reported by Lambert et al. (1999). Data of the reinforced soil constituents are: $E_c = 15000$ kPa; $E = 1500$ kPa; $H = 18$m; $\eta = 8.6\%$; $s = 3$ m; $Q = 910$ kN; $a = 0.5$ m; $b = 1.7$ m; and $k_h = 3.7\ 10^{-8}$ m.s^{-1}.

According to Figure 5.38, the predicted settlement is 25% higher than the measured settlement. However, the trend of the poroelastic model fits with measured settlement evolution. The difference in the amount of consolidation settlement may be attributed to the considered values of soil parameters (for example, the coefficient of horizontal consolidation of soft soil and geometrical parameters of the UCM).

Final Comments

It is important to emphasize that the poroelastic model presented above provides both the evolution of the degree of horizontal consolidation and the

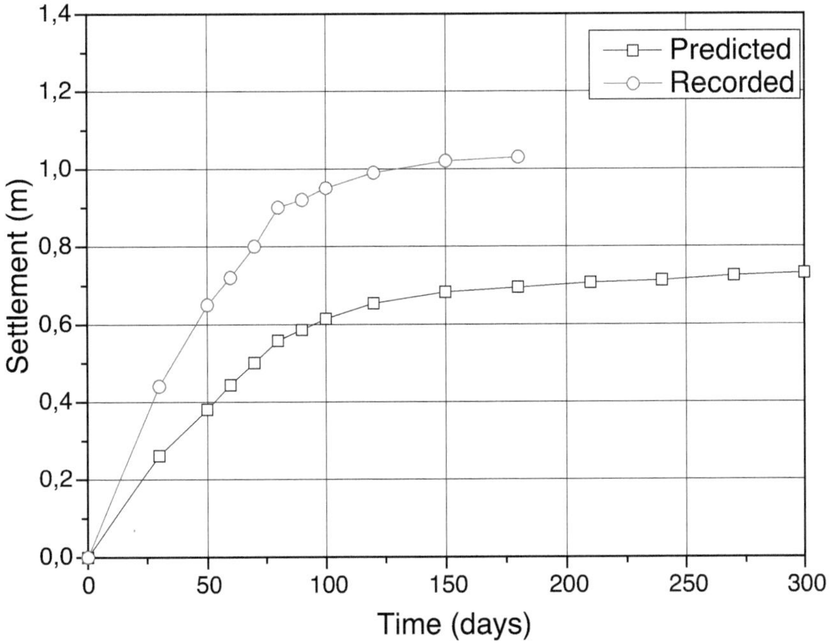

Figure 5.38 Evolution of predicted and measured settlement vs. time

related settlement for a given loading history under which the reinforced soil is subjected. It has been illustrated that such evolution obviously depends on the IAR.

Further, it has revealed that the coefficient of horizontal consolidation has significant influence on the predictions obtained from the poroelastic model. Because of unsatisfactory laboratory tests to determine this parameter, the calibration of predicted results by the poroelastic model becomes more challenging. Meanwhile, more investigations with field data should be considered to calibrate the proposed porous elastic model.

The predictions obtained by the poroelastic model, as detailed in this chapter, have been programmed in *Columns 1.01* software presented in Chapter 4.

5.5 CONCLUSIONS

The numerical study of the improvement of soft soil due to stone column installation induced by lateral expansion has been carried out by using the FE code Plaxis V9.2D.

The numerical dummy material procedure has been performed to simulate the lateral expansion of a single stone column and a group of stone columns in soft clay using two respective reinforced soil models in axisymmetric conditions. Improvement of soft soil properties has been evidenced as a result of the induced primary consolidation in post-installation of stone columns.

Numerical predictions carried out by a UCM indicates a better improvement in Young's modulus when the Mohr Coulomb constitutive law is adopted for soft clay, rather than the HSM. Therefore, much better reduction in settlement is predicted when an improved Young's modulus of soft clay is taken into account.

Investigation of an equivalent group of stone columns modeled in axisymmetric conditions led to a conclusion that significant improvement in Young's modulus is noted with respect to the adoption of the UCM. Importantly, an optimized IAR can be identified when the improvement in Young's modulus of soft soil is considered.

Laboratory experimental work was performed on remolded soft kaolin in order to estimate its improvement resulting from a laterally expanded stone column. Such reinforcement was simulated by a controlled lateral expansion applied within initially consolidated soft kaolin hollow specimens in isotropic conditions. Different rates of cavity expansion were tested up to twice the initial volume of the cavity. The lateral expansion was performed in drained conditions so that the soft kaolin would be completely consolidated under consolidation stress equal to 100, 200, and 300 kPa. All the hollow soft kaolin specimens were subjected to undrained shear loading until failure occurred with recorded excess pore pressure. The main findings that were a result of the experimental program are outlined below:

- The expansion of the cylindrical cavity improves the undrained Young's modulus and the undrained shear strength of the remolded soft kaolin.
- The increase in undrained shear strength essentially depends on the consolidation stress and the cavity expansion ratio. Such an increase is more pronounced at lower consolidation stress and higher cavity expansion ratio.
- The increase in undrained Young's modulus only depends on the rate of cavity expansion when it is less than 1.5. When the rate of cavity expansion exceeds 1.5, the improvement in undrained Young's modulus depends, in addition, on the stress of consolidation. If this latter increases, the improvement of Young's modulus decreases.
- As for the undrained Young's modulus, the ratio (E_u/c_u) increases when the ratio (V/V_0) is less than 1.5, especially for higher consolidation stresses. It is noted that the ratio (E_u/c_u) increases when the consolidation stress decreases.

- The dissipation of excess pore pressure after stone column installation turned out to be significant in enhancing the bearing capacity and reducing the settlement of reinforced soil.

The acceleration of primary consolidation of reinforced soil by drained material columns has been studied by means of a poroelastic model through which it is possible to characterize the settlement evolution of primary consolidation, with respect to time, for a given loading history or a specified stage construction. The evolution of consolidation settlement and the degree of horizontal consolidation, under a prescribed load exerted by the foundation, were calculated as functions of the IAR and linear elastic characteristics of the constituents of the UCM. The horizontal coefficient of consolidation appears to be the most important parameter that significantly affects the predictions by the suggested poroelastic model.

APPENDIX 5.1: BEHAVIOR OF NORMALLY CONSOLIDATED SOFT CLAYS

This section analyzes the behavior of normally consolidated soft clays. In some investigations the HSM has been considered to characterize the behavior of soft clays (Schanz et al., 1999, Guetif et al., 2006). Also the HSM was largely adopted to predict the behavior of sands.

After stress-strain curves that were recorded from drained triaxial tests performed on soft clays, it was shown that the consolidation pressure significantly affects the behavior in the range of small strains as well as large strains. As examples, the results obtained from experiments conducted by Biarez and Hicher (1994) and Tounekti et al. (2008) illustrated a strong nonlinear stiffness that depends on the stress level. This behavior cannot be reproduced if the Mohr Coulomb behavior model is envisaged for soft clays. Meanwhile the use of the HSM permits the prediction of an increase in stiffness as a result of effective stress consolidation (Schanz et al., 1999, Brinkgreve and Vermeer, 1998). Compared to the Mohr Coulomb model, the implementation of an elasto-plastic HSM provides the advantage in controlling the stress-stiffness dependency for a given loading path.

The stress-strain behavior for primary loading is characterized by the stiffness modulus E_{50}, which is considered instead of the initial tangent modulus E_i for small strain. For the latter, the experimental determination appears to be more difficult. The modulus E_{50} is determined from deviatoric stress-axial strain curves recorded after drained triaxial tests. E_{50}, which corresponds to half of the maximum shear strength, is calculated from the stress-stiffness relation (Schanz et al., 1999):

$$E_{50} = E_{50}^{ref}\left(\frac{\sigma_3 + C'\cot g\varphi'}{\sigma_3^{ref} + C'\cot g\varphi'}\right)^{\varpi} \tag{5.27}$$

The parameter ϖ is already introduced by Equation (5.1).

This HSM does not involve a unique relationship between the drained triaxial Young's modulus E_{50} and the oedometer Young's modulus E_{oed}. This latter is determined from:

$$E_{oed} = E_{oed}^{ref}\left(\frac{\sigma_1 + C'\cot g\varphi'}{\sigma_1^{ref} + C'\cot g\varphi'}\right)^{\varpi} \tag{5.28}$$

For unloading and reloading stress paths, the recognized modulus is (Figure 5.7):

$$E_{ur} = E_{ur}^{ref}\left(\frac{\sigma_3 + C'\cot g\varphi'}{\sigma_3^{ref} + C'\cot g\varphi'}\right)^{\varpi} \tag{5.29}$$

E_{oed}^{ref} is determined for a referenced major principal stress $\sigma_1^{ref} = 100$ kPa.

E_{50}^{ref}, E_{ur}^{ref} are determined for a referenced minor principal stress $\sigma_3^{ref} = 100$ kPa.

More details related to the HSM constitutive law can be found in Brinkgreve and Vermeer (1998) and Schanz et al. (1999).

In order to study the influence of stone column installation on the behavior of soft clay, parameters of the HSM are adopted in the numerical simulation (Biarez and Hicher, 1994). These parameters are calibrated with experimental data from oedometer and drained triaxial tests performed on normally consolidated clay.

Using the HSM parameters proposed in Table 5.3 for soft clay, the loading paths of oedometer and drained triaxial tests are effectively reproduced (Biarez and Hicher, 1994).

It was also reported that the HSM fairly models other loading paths such as the pressuremeter test (expansion of cylindrical cavity), Biarez et al. (1998), Brinkgreve and Vermeer (1998), Schanz et al. (1999).

The HSM allows the simulation of various loading paths, such as those observed in several soil mechanics in situ tests, for normally consolidated soft clays. For this reason the implementation of the HSM is intended to analyze,

Table 5.3 Hardening soil model parameters for clays (Guetif et al., 2008)

γ (kN/m³)	E_{50}^{ref} (kPa)	E_{ur}^{ref} (kPa)	E_{oed}^{ref} (kPa)	ϖ	C' (kPa)	φ' (°)	ψ (°)	K_0	k_h m/day
18 to 17	2500	20000	1300	1	2	25	0	0.7	2.10^{-04}

first, the soft clay behavior that results in pre- and post-stone column installation and the expected improvement in Young's modulus.

APPENDIX 5.2: DETAILS OF THE POROELASTIC MODEL

The solution of the poroelastic problem presented in Section 5.4.3 is detailed in Fessi-Guetif (2004). This appendix presents some functions and constants of the solution of the poroelastic problem. The expressions of constants $D_u(a)$ and $D_d(a)$ appearing in the differential Equation (5.18) are:

$$D_u(a)=\left[(a^2-b^2)\eta v_c E\right]\left(ab\left[(1-\eta)\lambda_c+2(\eta\lambda+(1+\eta)\mu)v_c\right]\right)^{-1} \quad (5.30)$$

$$D_d(a)=-\left[(a^2-b^2)\eta v_c(\lambda_c-\lambda)\right]\left(ab\left[(1-\eta)\lambda_c+2(\eta\lambda+(1+\eta)\mu)v_c\right]\right)^{-1} \quad (5.31)$$

The expressions of functions X and Z appearing in Equation (5.19) are:

$$X=abD_u(a) \quad (5.32)$$

$$Z=-\frac{(b^2-a^2)}{2}-abD_d(a) \quad (5.33)$$

The expressions of coefficients θ, χ, ω and ψ are detailed as follows:

$$\theta=\frac{abD_u(a)}{\left[(1-\eta)E-2(\lambda_c-\lambda)\frac{a}{b}D_u(a)\right]\pi b^2} \quad (5.34)$$

$$\chi=\frac{a\frac{K}{\gamma_w}E}{\left[(1-\eta)E-2(\lambda_c-\lambda)\frac{a}{b}D_u(a)\right]\pi b^2} \quad (5.35)$$

$$\omega=\frac{(b^2-a^2)}{2}+abD_d(a)-X\frac{\left[E_{oed,r}-2(\lambda_c-\lambda)\frac{a}{b}D_u(a)\right]}{\left[(1-\eta)E-2(\lambda_c-\lambda)\frac{a}{b}D_u(a)\right]} \quad (5.36)$$

$$\psi=a\frac{K}{\gamma_w}E\frac{\left[E_{oed,r}-2(\lambda_c-\lambda)\frac{a}{b}D_u(a)\right]}{\left[(1-\eta)E-2(\lambda_c-\lambda)\frac{a}{b}D_u(a)\right]} \quad (5.37)$$

where $E_{oed,r}$ denotes the homogenized oedometer modulus of reinforced soil defined by Equation (4.15).

REFERENCES

AFNOR, (French Standard) P 94-074 Sols: reconnaissance et essais. Essais à l'appareil triaxial de révolution. Appareillage, Préparation des éprouvettes et Essais (UU), (CU+u) et (CD); October 1994.

Alamgir, M. and Zaher, S. M., Field investigation on a soft ground of Bangladesh reinforced by granular piles. Proc. Int. Symp. Reinforcement Earth, Nov. 14–16th, 2001, Kyushu (Japan). Landmarks in earth Reinforcement, Ochiai et al. Editors, 2001, 517–522.

Al-Khafaji, Z. A. and Craig, W. H., Drainage and reinforcement of soft clay tank foundation by sand columns. Géotechnique, 50 (6), 2000, 709–713.

Almeida, M. S. S., Santa Maria, P. E. L., Martins, I. S. M., Spotti, A. P. and Coelho, L. B. M., Consolidation of a very soft clay with vertical drains. Géotechnique, 50 (6), 2000, 633–643.

Barron, A., Consolidation of fine grained soils by drains wells. American Society of Civil Engineers, Journal of Soil Mechanics, 73, SM6, 1947, 718–743.

Bergado, D. T., Rantucci, G. and Widodo, S., Full scale load test of granular piles and sand drains in the soft Bangkok clay. Proc. Symp. In situ reinforcement of soils and rocks, ENPC Press, Paris, 9–11 October. Session 1b, Foundations: stone inclusions, 1984, 111–118.

Biarez, J. and Hicher, P. Y., *Elementary Mechanics of Soil Behaviour Saturated Remoulded Soils*, 1994. Balkema, Rotterdam.

Biarez, J., Gambin, M., Gomes-Corriea, A., Flavigny, E. and Branque, D., Using pressumeter to obtain parameters of elastoplastic models for sands. Proc. 1st Int. Conf. on Site Characterization. ISC'98, Atlanta, Georgia, USA, 1998, 747–752.

Bouassida, M., Frikha, W. and Canou, J., Experimental study of stone columns installation in kaolin clay. Proc. Int. Conf. on Ground Improvement and Ground Control, October 30–November 2, 2012. University of Wollongong (Australia), Indraratna et al. (Eds), 1, 763–768.

Bouassida, M., Design of columnar-reinforced foundations. Invited paper, Proc. Int. Conf. on Ground Improvement and Ground Control, October 30–November 2, 2012. University of Wollongong (Australia), Indraratna et al. (Eds), 1, 599–608.

Brinkgreve, R. B. T. and Vermeer, P. A., Plaxis-Finite Element Code for Soil and Rocks Analysis. Version 8, 1998. AA. Balkema, Rotterdam Brookfield.

Callanan, J. F., Design and construction monitoring of stone column foundations in hydraulic fill. Proc 10th European Conference on Soil Mechanics and Foundation Engineering, Florence, May 26–30, V1, Pub. Rotterdam, Balkema, 1991, 349–352.

Corneille, S., Masrouri, F. and Soyez, B., Essai de chargement en grandeur réelle d'une colonne ballastée isolée—Proc. ELU-ULS. Marne-la-Vallée, Editions du LCPC, 2006, 71–78.

Coussy, O., *Porous Media mechanics (in French).* Edit. Technip, 1991, 1–133. Paris.

Debats, J. M., Savasta, P. and Gambin, M. P., Vibro-replacement and static pre-loading of fine-grained granular soils: comparison and importance of in situ control testing methods (in French) 12th Europ. Conf. Soil Mech. & Geotech. Eng. Amsterdam, 1999 Barends et al. Eds, (3), 1761–1764.

Debats, J. M., Guetif, Z. and Bouassida, M., Soft soil improvement due to vibro-compacted columns installation. Proc. International Workshop on Geotechnics of soft soils. Theory and practice, Sept. 17–19, Noordwijkerhout (Netherlands), 2003, Vermeer et al. Eds, 551–556.

Dormieux, L. and Bourgeois, E., *Introduction to the micromechanics of porous media.* Edit. Broché, 2002, Paris.

Ellouze, S., Bouassida, M., Debats, J. M. and Scharff, G., Simulation of the installation of a group of columns in soft clay. Proc. Int. Conf. Geotech. Eng. "Innovative Geotechnical Engineering" March 24–26th, Hammamet (Tunisia), Bouassida, et al. (Eds), 2010, 95–103.

Ellouze, S. and Bouassida, M., Prediction of the settlement of reinforced soft soil clay by a group of stone columns. Proc. 2nd Int. Conf. New Develop. in Soil Mech. Geotech. Eng., NEU Nicosia, 2009, 182–187.

Ellouze Zribi, S., Improved soil characteristics for settlement prediction of reinforced soil by stone columns. PhD's dissertation (in French), defended on November 10, 2010. National Eng. Sch. of Tunis, Tunisia.

Fessi Guetif, Z., On the prediction of settlement of foundations on reinforced soil by columns. PhD's dissertation (in French), defended on March 5, 2004. National Eng. Sch. of Tunis, Tunisia.

Frikha, W., Bouassida, M. and Canou, J., Calibration of an elastoplastic model for the prediction of stone column ultimate bearing capacity. Proceedings of selected sessions of Geo Congress 2008. Geosustainability and Geohazard mitigation. Special Publication ASCE GEO-Institute (invited paper), Edit. K. R. Reddy et al. 604–611.

———. Observed behaviour of laterally expanded stone column in soft soil. Geotech. Geol. Eng. J., 31 (2), 2013, 739–752.

Gambin, M. P., Puits ballastés à la Seyne-sur-Mer. Proc. Int. Symp.: Renforcement en place des sols et des roches. Paris, 1, 1984, 139–145.

Guetif Fessi, Z. and Bouassida, M., Analytical estimate of settlement evolution of soft soil reinforced by stone columns. Proc. 16th Int. Conf. of Soil Mech. and Geotech. Eng. September 12th–16th, 2005, Osaka, Japan, 3:1355–1358.

Guetif, Z., Bouassida, M. and de Buhan, P., Prévision du tassement d'un sol renforcé par colonnes à l'aide d'un modèle poro-élastique. Comptes rendus Symposium International sur l'amélioration des sols en place (ASEP-GI). 9–10th September 2004, 155–162.

Guetif, Z., Bouassida, M. and Tounekti, F., Numerical simulation of stone column installation using advanced elastoplastic model for soft soil. New Horizons in Earth Reinforcement—Otani, Miyata & Mukunoki (eds). 2008, Taylor & Francis Group, London, 441–446.

———. Numerical simulation of vibro-compacted stone column installed in Tunis soft clay. Proc. 1st Euro Mediterranean Symposium in Advances on Geomat. and Struct. May 3rd–5th, Tunisia, 2006, 233–238.

Guetif, Z., Bouassida, M. and Debats, J. M., Parametric study of the improvement due to vibrocompacted columns installation in soft soils. Proc.13th Reg. African Conf. of Soil Mech. & Geotech. Eng., Marrakech (Morocco) December 8–11th, 2003, 463–466.

———. Improved soft clay characteristics due to stone column installation. *Computers and Geotechnics*. 34 (2), 2007, 104–111.

Kirsch, F., Vibro Stone Column Installation and its Effect on Ground Improvement. 2005, www.researchgate.net/publication/242141654.

Lambert, S., Naidji, T. and Berthelot, P., Amélioration des sols limoneux vasards par colonnes ballastées et préchargement statique. Proc. 15th Int. Conf. Soil Mech. Geot. Eng. Istanbul 27–31 August, 2001, (3), 1779–1784.

Naama Engineering & Consulting. Geotechnical report of project "Damiette LNG tanks." 2001, Vibroflotation Europe.

Salençon, J., Visoélasticité, 1983, Eds. Presses de l'École Nationale des Ponts et Chaussées, 1983, Paris.

Schanz, T., Vermeer, P. A. and Bonnier, P. G., 1999. The hardening soil model: Formulation and verification. Beyond 2000 in Computational Geotechnics 10 years of Plaxis. 1–16. Balkema, Rotterdam.

Tounekti, F., Bouassida, M., Klai, M. and Marzougi, I., Etude expérimentale en vue d'un modèle de comportement pour la vase de Tunis. *Revue Française de Géotechnique*. 2008, 122(1), 25–36.

Vautrain, J., Comportement et dimensionnement des colonnes ballastées. *Revue Française de Géotechnique*, 11(1), 1980, 59–73.

6

BEHAVIOR OF FOUNDATIONS ON SOILS REINFORCED BY COLUMNS

ABSTRACT

On the basis of a preliminary design using an optimized improvement area ratio (IAR) suggested by the methodology detailed in Chapter 4, the prediction of the behavior of foundations on soil reinforced by columns is analyzed by implementing numerical computations carried out by the Plaxis V9.2D and FLAC3D codes. Three Tunisian case histories—two oil tanks and an LPG (liquefied petroleum gas) storage facility—are investigated.

When considering the linear elastic behavior for constituents of the reinforced soil, it is found that the predictions of settlements using the *Columns 1.01* software are in very good agreement with numerical results obtained both by the Plaxis code, in plane strain and axisymmetric conditions, and by the FLAC3D code.

By adopting the Mohr-Coulomb constitutive law for columns material and the hardening soil model (HSM) for soft clay, the predicted variation of long-term settlement in plane strain and axisymmetric conditions was found to be quite similar. The acceleration of consolidation of compressible soft clay due to the enhanced drainage property of granular column material has been proven.

Further, the reinforcement by floating columns represents a suitable option of foundation as validated by a Tunisian case history when the evolution of long-term settlement of unreinforced sublayers is admissible.

6.1 INTRODUCTION

Since the last decade, the study of the behavior of foundations on soils reinforced by columns has known an increasing number of contributions. As for other geotechnical applications, the development of finite element and finite difference codes, in conjunction with advanced constitutive laws for soils, enhanced the accuracy of numerical computations using either 2D or 3D modeling of reinforced soils by a group of columns.

Among several contributions, Han and Ye (2001) demonstrated the stress transfer from the soil to stone columns and dissipation of excess pore water pressures due to drainage and vertical stress reduction during the consolidation. Six et al. (2012) studied the effect of horizontal stress in reinforced clay by expanded stone columns. Those authors carried out finite element calculations to predict stone columns behavior using analytical methods and conventional engineering methods. It was found that the at-rest earth pressure coefficient K_0 is related to the initial state of stresses caused by the compression of in situ soil due to the installation of column material. The importance of defining the initial state of stress due to stone columns installation is of big importance for the design of a stone column foundation.

Sexton and McCabe (2013) evaluated the settlement improvement factors for different area-replacement ratios, modular ratios, and column lengths. The primary settlement improvement factors are in good agreement with some of the more popular analytical design methods, while the creep-settlement improvement factors are either equivalent or lower, depending on the approach used. The primary settlement improvement factors show a dependence on the modular ratio, whereas it appears that the corresponding creep-settlement improvement factors are relatively independent of it. Chai et al. (2015) simulated the behavior of a test embankment constructed on a soft clay deposit in Saga, Japan, by 3D and 2D finite element analyses (FEA). Floating soil–cement columns had been installed in the clay prior to construction of the embankment. A comparison of the results of 3D and 2D FEA indicates that the 2D analysis incorrectly predicted the lateral displacement and bending moment in the columns under the toe of the embankment. In the 2D analysis, the rows of columns were modeled by continuous walls that partially block the interaction between the soil layers and the columns, thus influencing the simulated lateral displacement and bending moment in the column. It has been postulated that compaction of fill material during the construction process has a significant influence on both the magnitude and pattern of the lateral displacement of the column under the toe of the embankment. The measurements and FEA results indicate that the columns not only reduced the total settlement but also accelerated the settlement rate of the deposit under the embankment loading, due to the much higher stiffness of the column material.

Indraratna et al. (2013) recently proposed a novel finite-difference model to analyze the response of stone column–reinforced soft soil under embankment loading, adopting the free strain approach, and considering both arching and clogging effects. The proposed model is validated by comparing it with existing models and field data that indicate the suitability and accuracy of the solutions.

In this chapter, three main goals are targeted:

- The first objective is to discuss the usefulness of linear elastic behavior to predict the settlement of reinforced soil by columns. Accordingly, the predictions by *Columns 1.01* software are compared with numerical results predicted by the unit cell model (UCM) and group of columns modeling in linear elasticity.
- The second objective, regardless of the consolidation aspect, is to compare the difference between settlement predictions in linear elasticity by *Columns 1.01* software and the elastoplastic Mohr-Coulomb model using Plaxis V9.2D and FLAC3D codes.
- The third objective is to study the behavior of soft soils reinforced by floating stone columns by taking account of the primary consolidation (Ellouze and Bouassida, 2009). The follow-up of structures founded on reinforced soil by columns is illustrated by a Tunisian case history. Main findings are synthetized in the conclusions' part. Finally, forthcoming developments on soils reinforced by columns are outlined.

6.2 STUDY OF REINFORCED SOIL BEHAVIOR: SELECTED CASE HISTORIES

6.2.1 Foundations on Compressible Clays Reinforced by End-bearing Columns

In the following section, two Tunisian study cases are considered to carry out numerical computations using the FLAC3D code. The first case study was considered by Bouassida and Hazzar (2012)—"Oil tank on reinforced soil by stone columns (Zarzis, Tunisia)." The second case study is the "Oil tank on end-bearing sand compaction columns foundation (La Goulette)" that has already been investigated in Chapter 4 (§4.5.1).

6.2.1.1 Oil Storage Tank at Zarzis (Tunisia)

This case study was considered by Bouassida and Hazzar (2012) to verify the executed design of a reinforced loose silt sand layer by end-bearing stone columns resting on marl stratum. The applied load (120 kPa) by the oil tank with a diameter of 54 m exceeds the allowable bearing capacity of the tank foundation. Performing the methodology of design detailed in Chapter 4, it was revealed that a minimum IAR of 13% is necessary to reinforce the loose silt sand layer. Meanwhile, the allowable settlement of reinforced soil was limited to 6 cm. The subsequent finding was that an optimized IAR of 30.64% is suitable to ensure the stability of the oil tank. However, a very conservative reinforcement was executed with an IAR equal to 35% that consisted in the installation of 700 stone columns with a diameter of 1.2 m, in a triangular pattern, with axis to axis spacing of 2.06 m.

Figure 6.1 presents the FLAC3D numerical model composed of a central row crossing the tank center with the dimensions: 55.1 m in length along x direction, 3.8 m in width along y direction, and 7 m in depth as the length of end-bearing stone columns. This reinforced soil model comprising 29 columns is modeled by a generated mesh constituted by 6,496 finite difference zones and 7,965 grid-points. The reinforced soil model subjected to the tank load of 120 kPa has the following boundary conditions: horizontal displacement along x and y directions is assumed zero and the horizontal and vertical displacements are zero at the stratum depth level. Figure 6.2 represents a zoomed view of the reinforced soil model in Figure 6.1, showing the stone columns installed in triangular patterns.

Computations were carried out by adopting Mohr-Coulomb elastoplastic behavior without taking account of the primary consolidation option. The geotechnical parameters of an adopted numerical model are displayed in Table 6.1.

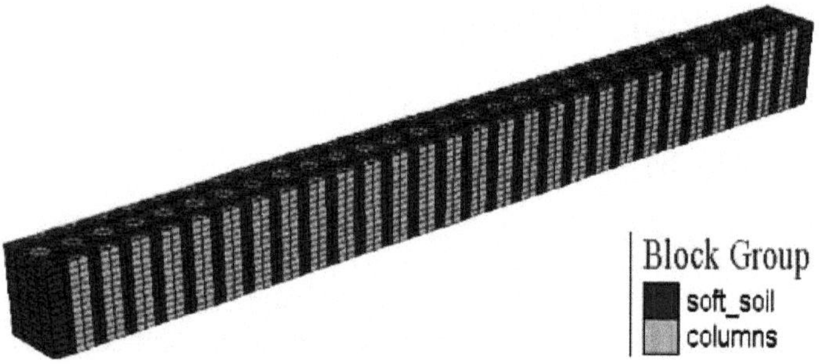

Figure 6.1 Geometry of FLAC3D model of reinforced ground by stone columns (Zarzis case study)

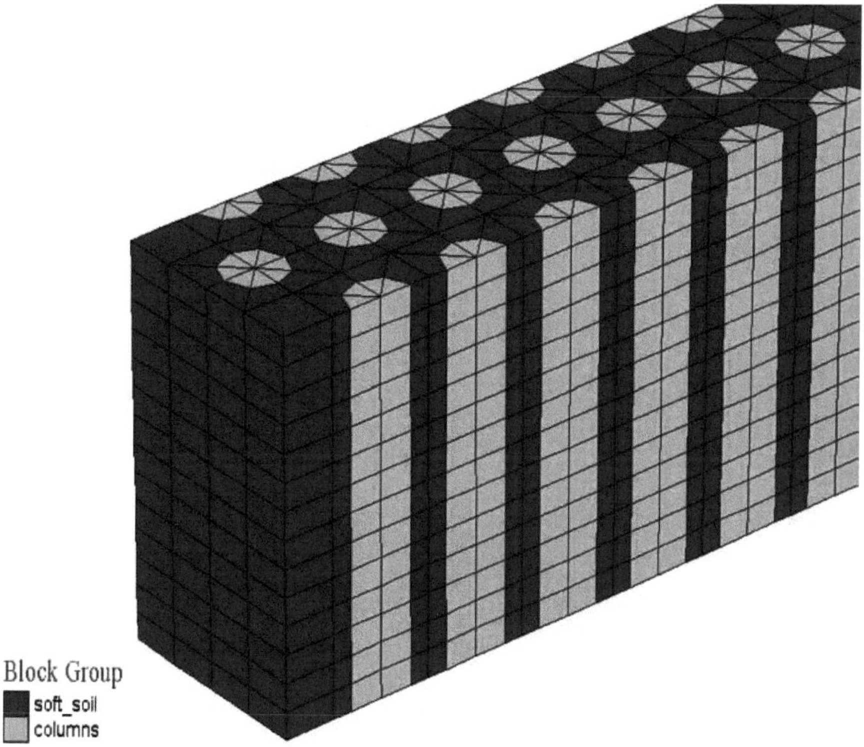

Figure 6.2 Zoomed in view of columns pattern and generated mesh by FLAC3D code of reinforced ground (Zarzis case study)

Obtained Results and Interpretations

Figure 6.3 shows the iso-values of vertical displacement within the unreinforced ground as computed by the FLAC3D code. The evolution of vertical displacement of unreinforced soil at the centerline of the tank versus the number of iterations is plotted in Figure 6.4 and indicates a maximum vertical displacement of 23.7 cm. Further, the iso-values of vertical displacement within the reinforced ground by stone columns appear in Figure 6.5. The evolution of vertical displacement of reinforced soil versus the number of computation iterations at the centerline of the tank is plotted in Figure 6.6 and indicates a maximum vertical displacement of 5.2 cm.

Figure 6.5 well illustrates the uniformity of predicted settlement at the surface of the reinforced soil as shown by the contour of iso-values of vertical displacement in the central part of the tank. This result confirms the validity of assumed equal settlement at the surface of reinforced ground, as considered in Chapter 4. At the columns' head and the initial soil in between, the magnitude

Table 6.1 Geotechnical parameters of the elastoplastic model of reinforced soil in FLAC3D (Zarzis case study)

Parameters	Thickness or length (m)	Young's modulus (MPa)	Poisson's ratio	Friction angle (°)	Cohesion (kPa)	Total unit weight (kN/m³)
Columns material	7	36	0.33	42	0	18
Loose silt sand	7	3.6	0.33	0	25	17

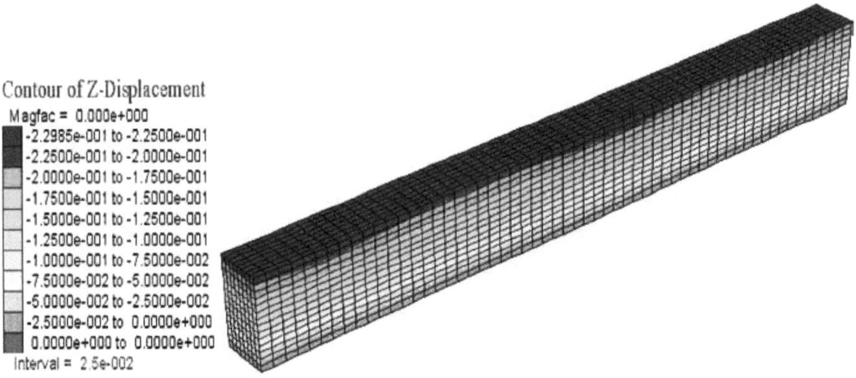

Figure 6.3 Iso-values of vertical displacement within the unreinforced ground (Zarzis case study)

of vertical displacement is quite similar. The settlement reduction factor varies from 4.0 to 4.56—this demonstrates the efficient role of stone columns in settlement reduction.

Table 6.2 summarizes the predictions of maximum settlement as obtained by the FLAC3D code and by the methods incorporated in *Columns 1.01* software (Bouassida and Hazzar, 2012), both for the unreinforced and reinforced soil cases. All those methods consider the linear elastic behavior for the initial soil and column material. It is noted that the settlement predictions by the FLAC3D code are quite close to those obtained by the analytical method proposed by Bouassida et al. (2003), which uses the group of columns model, similar to the FLAC3D code. The relative difference between these predictions is 1.8% and 11.5% for the unreinforced soil and reinforced soil respectively, which is quite negligible. This first case study confirms the usefulness of analytical methods of settlement predictions programmed within the *Columns 1.01* software that

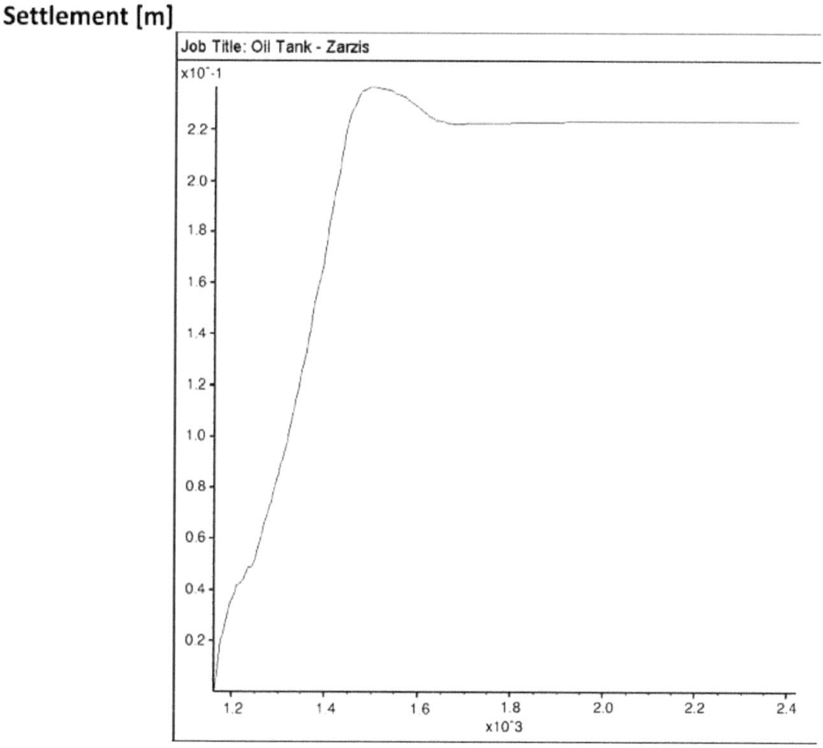

Figure 6.4 Evolution of vertical displacement of unreinforced soil at centerline of tank vs. number of iterations

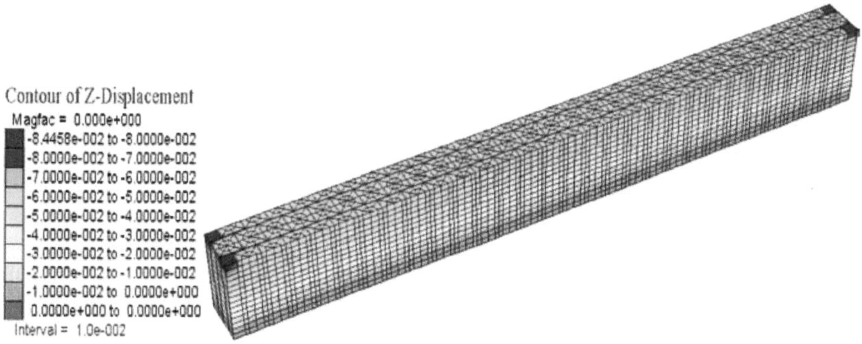

Figure 6.5 Iso-values of vertical displacement within the reinforced ground (Zarzis case study)

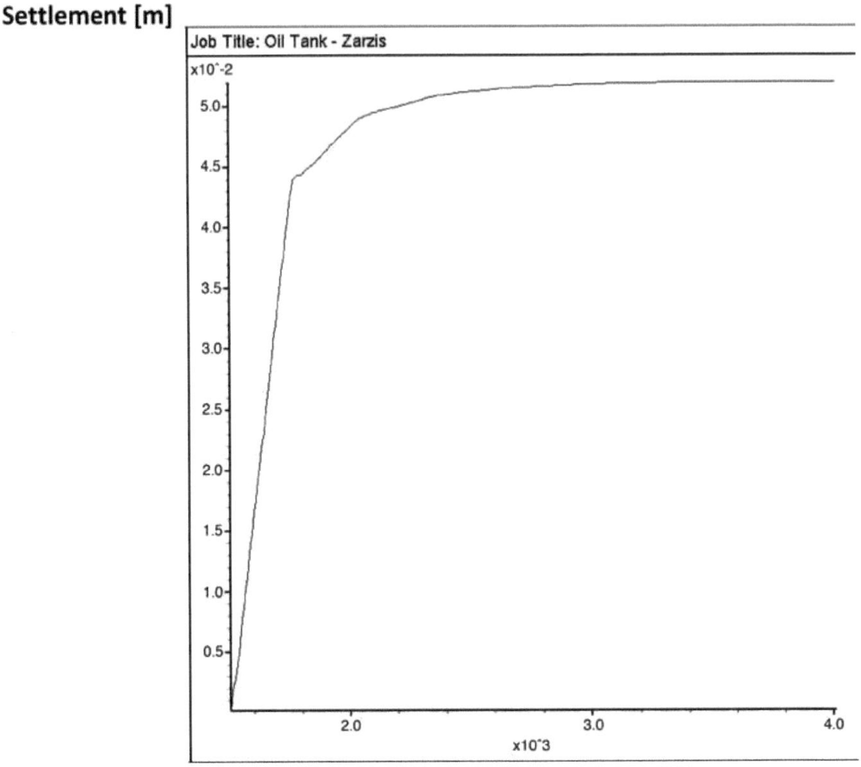

Figure 6.6 Evolution of vertical displacement of reinforced soil at centerline of tank vs. number of iterations

Table 6.2 Settlement predictions at centerline of tank (Zarzis case study)

Prediction method		FLAC 3D code	Bouassida et al. (2003)	French method (2005)	Balaam and Booker (1981)
Settlement (cm)	Unreinforced soil	23.7	23.28	—	—
	Reinforced soil	5.2	5.8	5.5	5.1

provides the optimized IAR to be considered in numerical computations for the study of the behavior of reinforced soil by columns.

6.2.1.2 Oil Storage Tank at La Goulette (Tunisia)

Data from this case study are detailed in Chapter 4 (§4.5.1). The reinforcement by end-bearing sand columns on a soft clay layer of 10 m thickness has been considered for an applied surcharge load of 80 kPa by a circular tank of 30 m in diameter. Considering the allowable settlement of reinforced soil equals 10 cm, the optimized IAR, as determined by the *Columns 1.01* software, is 29.95%. That was executed by 729 sand columns of 0.6 m in diameter with axis-to-axis spacing of 1.06 m installed in a triangular pattern.

Figure 6.7 depicts the corresponding FLAC3D numerical model of reinforced soil that consists of a central row crossing the tank center with dimensions of 30.74 m in length along the x direction, 2.12 m in width along the y direction, and 10 m in depth as the length of the end-bearing sand columns. This reinforced soil, comprising 29 columns, is modeled by a generated mesh constituted by 18,320 finite difference zones and 16,947 grid-points. The reinforced soil model subjected to the tank load of 80 kPa has the following boundary conditions: horizontal displacement along x and y directions is assumed zero and the horizontal and vertical displacements are zero at the stratum depth level.

Computations are carried out by adopting Mohr-Coulomb elastoplastic behavior without taking account of the primary consolidation option. The geotechnical parameters of the numerical model are given in Table 6.3.

Obtained Results and Interpretations

Figure 6.8 shows the iso-values of vertical displacement within the unreinforced ground as computed by the FLAC3D code. The predicted maximum vertical

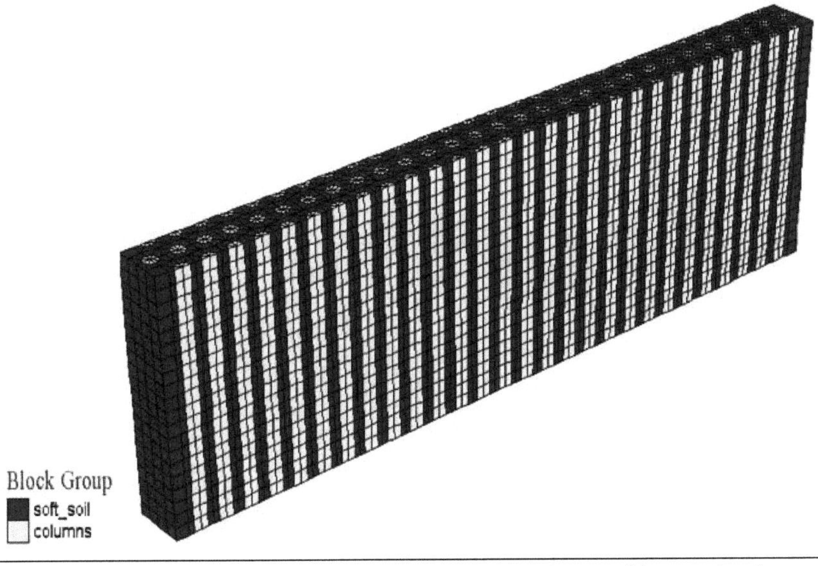

Figure 6.7 Geometry of FLAC3D model of reinforced ground by sand columns (La Goulette case study)

Table 6.3 Geotechnical parameters of the elastoplastic model of reinforced soil in FLAC3D (La Goulette case study)

Parameters	Thickness or length (m)	Young's modulus (MPa)	Poisson's ratio	Friction angle (°)	Cohesion (kPa)	Total unit weight (kN/m^3)
Sand (columns)	10	20	0.33	38	0	18
Tunis soft clay	10	2	0.4	0	24	17

displacement at the surface of the unreinforced centerline of the tank is 20.4 cm. Further, the predicted maximum vertical displacement at the surface of the reinforced ground is 9.9 cm (Figure 6.9).

Table 6.4 summarizes the predictions of settlement as obtained by the FLAC3D code and by the methods incorporated in *Columns 1.01* software (Bouassida and Hazzar, 2012) that consider the linear elastic behavior for the initial soil and column material as well. It is noted that the settlement prediction by FLAC3D code is close to the one obtained by the analytical method proposed by Bouassida et al. (2003) which uses the group of columns model similar to the FLAC3D code. The relative difference between the predicted settlements of reinforced soil is less than 1.2%, which is quite negligible.

Conclusions

The settlement of the tank foundation on reinforced soil by end-bearing columns has been predicted by a numerical model using the FLAC3D code. The

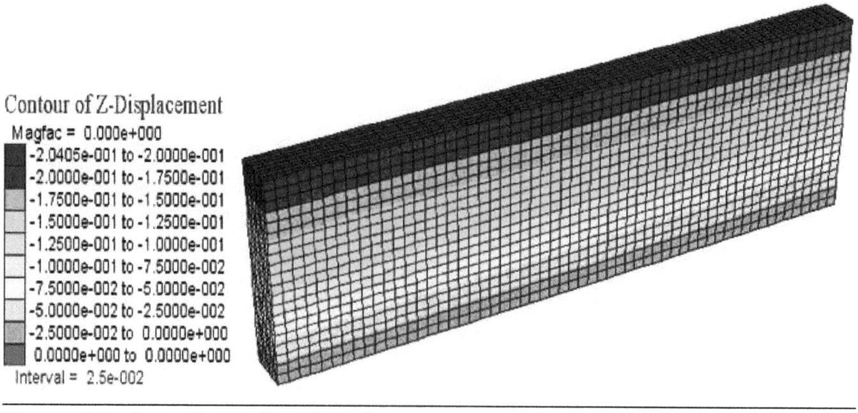

Figure 6.8 Iso-values of vertical displacement within the unreinforced ground (La Goulette case study)

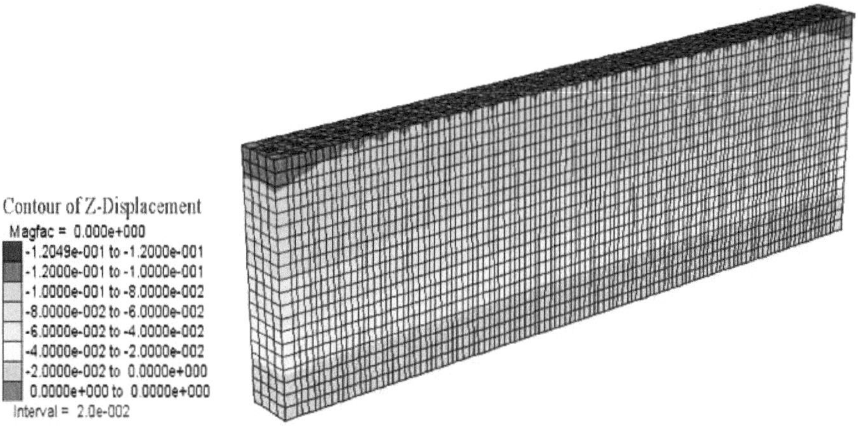

Figure 6.9 Iso-values of vertical displacement within the reinforced ground (La Goulette case study)

Table 6.4 Settlement predictions at centerline of oil tank (La Goulette case study)

Prediction method		FLAC 3D code	Bouassida et al. (2003)	French method (2005)	Balaam and Booker (1981)
Settlement (cm)	Unreinforced soil	20.4	38.73	—	—
	Reinforced soil	9.88	10	8.6	7.48

analysis of two Tunisian case studies of oil tanks built on reinforced soil by end-bearing columns has been carried out by adopting an optimized IAR as determined by the software *Columns 1.01*. The following interpretations are drawn:

- The settlements of the unreinforced soil and reinforced soil were predicted using FLAC3D modeling composed of a central row crossing the tank center with appropriated dimensions and boundary conditions subjected to the same admissible applied load. Those predictions permitted the quantification of the settlement reduction factors of 4 and 2.1 for the case studies of Zarzis and La Goulette, respectively.
- Considering the elastoplastic behavior (without the consolidation option) for constituents of reinforced soil by columns, the numerical predictions by the FLAC3D code have confirmed the results provided by the methodology of design of column-reinforced foundations embodied in the *Columns 1.01* software. This finding confirms the efficiency of the methodology of design detailed in Chapter 4. Although the elastoplastic behavior for constituents of reinforced soil by columns was assumed by the numerical FLAC3D code, the related results were found identical

to those predicted by the new methodology of design that considers the linear elastic behavior for constituents of reinforced soil (Bouassida and Carter, 2014). Such results can be explained by the fact that settlement prediction is essentially foreseen for admissible loads that do not induce large deformation within the reinforced soil. It follows that the predictions by linear elastic and elastoplastic constitutive laws revealed quite similar outcomes. Indeed, the normalized admissible settlements for the two case studies were equal to 1/900 and 1/300 respectively, and show the settlement predictions were made in the range of small deformations. Hence, for preliminary design, the linear elastic behavior can be adopted as constitutive law for the prediction of settlement on the basis of an optimized IAR of foundations on reinforced soil by columns. Then, the study of behavior can be investigated by using a finite element or a finite difference code to focus on the stresses and displacements distributions within the reinforced soil.

6.2.2 Comparison Between Numerical Predictions in Plane Strain and Axisymmetric Analyses

The plane strain condition can suitably be adopted for structures having a current cross section with a lengthy extension as compared to the dimensions of the cross section of studied structure. The lineal loading condition of such structures shows that the displacement in the perpendicular direction to the cross section of the considered structure is negligible. In turn, axisymmetric analysis applies for structures having geometrical and loading symmetries with respect to a revolution axis. The Plaxis computations have been carried out by using generated mesh composed of fifteen nodes of triangular finite elements.

The case study of an oil tank at La Goulette (Tunisia) is now used to compare the difference between the predicted behavior of unreinforced and reinforced Tunis soft clay by sand columns, both in plane strain and axisymmetric conditions. This numerical study has been conducted by the Plaxis V9.2D finite element code by adopting the HSM as constitutive law for the Tunis soft clay. In fact, the simulation of observed behavior of Tunis soft clay specimens subjected to oedometer and triaxial tests showed that the HSM predictions are in good accordance with measured data, rather than predicted results obtained by the modified Cam-Clay model (Klai, 2014).

1st Computation Option: Plane Strain Condition

Performing the plane strain computation passes, first, by the determination of the equivalent width, B, of a squared tank that has the same area as the circular tank of radius $R = 15$ m, then $B=\sqrt{\pi}R=26.587m$ (Bouassida et al. 2015).

Figure 6.10 shows a 3D view of the La Goulette oil tank for which the plane strain condition is adopted in comparing between the 2D numerical results and the predictions obtained by the axisymmetric study that is the more realistic due to the cylindrical symmetry of the tank and applied load. The corresponding plane strain modeling is sketched in Figure 6.11.

6.2.2.1 Predictions by the HSM (Plane Strain Condition)

The evolution of the settlement of unreinforced soil at the edge of the oil tank sketched in Figure 6.12 shows a maximum consolidation settlement of 4.2 cm. The predicted maximum consolidation settlement is 21.8 cm at the centerline of the oil tank as consolidation time approaches ten years. It is clearly seen that the long-term settlement due to the consolidation of the soft clay layer slightly exceeds the settlement of 20.8 cm predicted by different methods that adopt linear elastic behavior (Table 6.4).

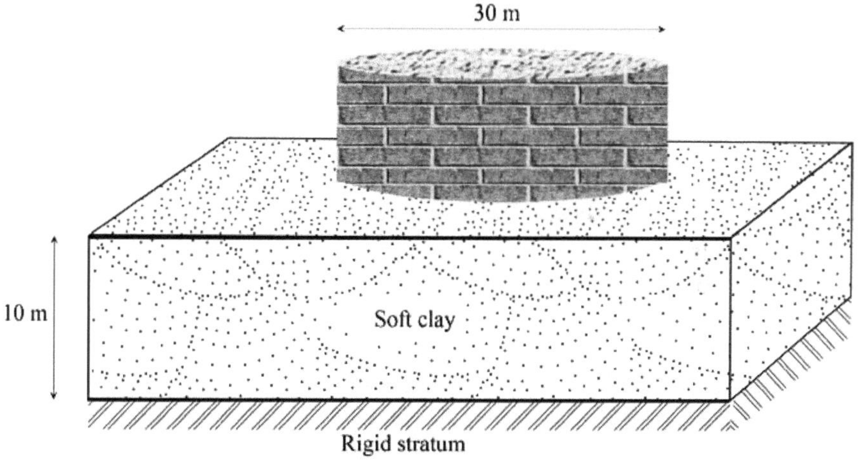

Figure 6.10 Three-dimensional view of oil tank at La Goulette site

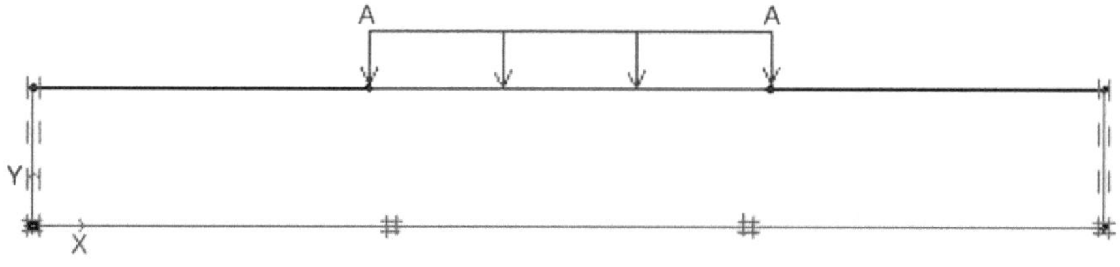

Figure 6.11 Plane strain numerical modeling (La Goulette case study)

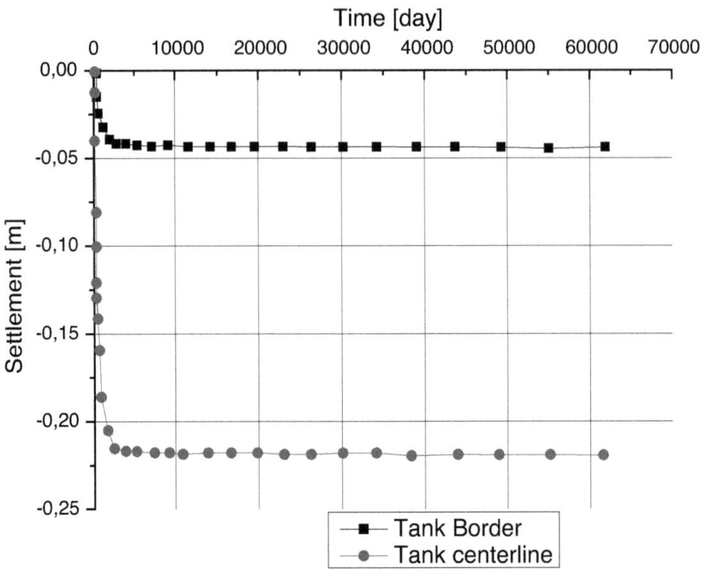

Figure 6.12 Predicted evolutions of consolidation settlement of unreinforced soil (plane strain condition)

In addition to the significant consolidation settlement predicted at the centerline of the oil tank, the differential settlement (difference between predicted settlements at the centerline and border of the oil tank) is not acceptable for its stability. The normalized differential settlement is approximately 0.3%, which notably exceeds the allowable limit of 0.1%. Therefore, to avoid unallowable settlements that compromise the stability of the tank, the reinforcement of the soft clay layer by columns is strongly recommended to reduce the total settlement and to accelerate the consolidation settlement by choosing a drained column material (like sand) with enhanced permeability.

The prediction of the behavior of a soft clay layer reinforced by sand columns was carried out in plane strain condition using the Plaxis V9.2D code. Therefore, the equivalence between circular and squared foundations on reinforced soil having the same optimized IAR (η_{opt} = 29.95%) permits the determination of equivalent width of reinforcing trenches (Figure 6.13). Consider the total number of sand columns N_c is kept identical for the circular and squared rafted foundations, hence the following relationship is obtained:

$$\eta_{opt}\pi R^2 = N_c \pi a^2$$

R and a are the radii of the circular rafted foundation and reinforcing columns, respectively. The total number of reinforcing columns, N_c, is then deduced: $N_c = \eta_{opt}(R/a)^2$.

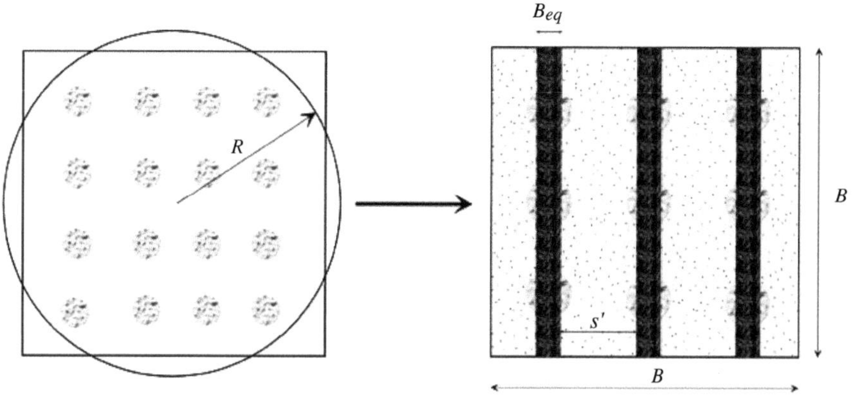

Figure 6.13 Equivalent plane strain model of circular tank resting on reinforced soil by sand columns (La Goulette case study)

From Figure 6.13 the width of reinforcing trench B_{eq} parallel to the width B of the equivalent squared foundation is:

$$B_{eq} = \eta_{opt} B / \sqrt{N_c} = 0.4795m$$

The spacing between edges of trenches s' is then deduced:

$$B = \sqrt{N_c}.B_{eq} + \sqrt{N_c}.s'$$

$$s' = B / \sqrt{N_c} - B_{eq}, \text{ then: } s' = 0.6805m$$

The perfect elastoplastic Mohr-Coulomb model has been adopted for the sand as constitutive columns' material with geotechnical properties displayed in Table 6.5.

Figure 6.14 shows the plane strain numerical model of Tunis soft clay reinforced by sand trenches subjected to the oil tank loading. At the surface of

Table 6.5 Geotechnical parameters of Mohr-Coulomb model adopted for sand columns material

γ_c [kN/m³]	18
e	0.5
E_c^* [kPa]	20,000
G [kPa]	7,519
ν_c	0.33
$E_{oed,c}$ [kPa]	29,630
φ_c [°]	38.0
$k_h = k_v$ [m/day]	10

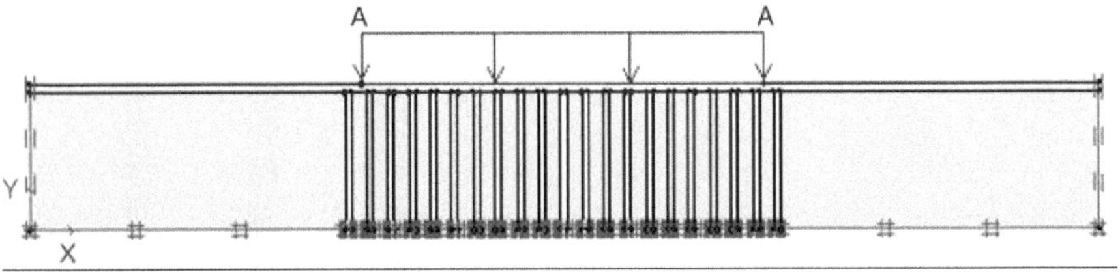

Figure 6.14 Plaxis plane strain model of Tunis soft clay reinforced by sand trenches subjected to oil tank loading (La Goulette case study)

reinforced soil a thin drainage layer of a thickness of 0.6 m is placed to permit the evacuation of water from the sand columns, which behave like vertical drains during the consolidation of soft clay. The variation of consolidation settlement of reinforced soil, in plane strain condition, is plotted in Figure 6.15, both at the centerline and at the border of the oil tank where the predicted values are 7 cm and 3.8 cm, respectively. The quasi-constant distribution of settlement at the surface of reinforced ground is owed to the presence of a drainage sand layer. Note that the variation of settlement stabilizes almost within 18 months, which shows the acceleration of consolidation of soft clay by sand columns. Take into account that the differential settlement equals the difference between predicted settlements at the centerline and the border of the oil tank. When comparing between Figures 6.12 and 6.15, note that the reinforcement by sand columns greatly reduces the differential settlement from 17.7 to 3.2 cm, which is an important benefit for the long-term stability of the oil tank.

6.2.2.2 Predictions by the HSM (Axisymmetric Condition)

The predictions of the behavior of unreinforced soil and reinforced soil by sand columns were carried out in axisymmetric condition. Numerical computations were run with Plaxis V9.2D code. The adopted parameters of the HSM for Tunis soft clay are given in Table 6.6.

The Plaxis' axisymmetric model of the La Goulette oil tank and its associated deformed mesh are shown in Figures 6.16 and 6.17, respectively. Figure 6.17 also illustrates the profile of consolidation settlement at the surface of the unreinforced soil. The evolution of consolidation settlement versus time at the surface of the unreinforced soil is sketched in Figure 6.18. It shows the primary consolidation is expected to end within 12.5 years. The maximum consolidation settlement of unreinforced soil at the centerline of the tank equals 26 cm, which exceeds the predicted consolidation settlement (22 cm) in plane strain condition shown in Figure 6.12. The same remark holds for the predicted settlements at the border of the tank, which are 8.3 cm and 4.2 cm in axisymmetric and

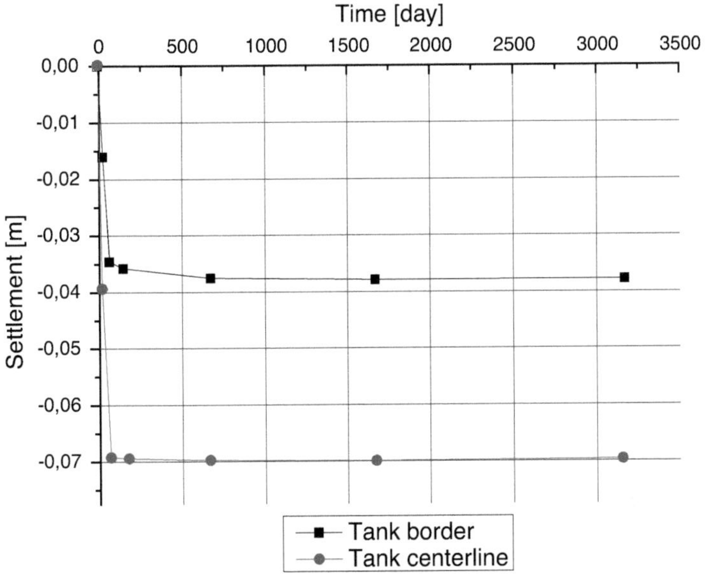

Figure 6.15 Predicted variations of consolidation settlement of reinforced soft clay by sand columns (plane strain condition)

Table 6.6 Geotechnical parameters of the HSM adopted for Tunis soft clay. E_{oed}^{ref} = 1,664 kPa; E_{50}^{ref} = 1,332 kPa; E_{ur}^{ref} = 9,892 kPa; P^{ref} = 100 kPa; R_f = 0.9.

γ_{sat}	[kN/m^3]	17
$k_h = k_v$	[m/day]	1.7410^{-4}
e_{init}	[–]	1.2
ϖ	[–]	1
C'	[kPa]	4.4
φ'	[°]	18
ψ	[°]	0
K_0	[–]	0.69

plane strain conditions, respectively. However, when adopting the consolidation procedure, it was found that the end of primary consolidation settlement remains unchanged (12.5 years) both for the plane strain and axisymmetric analyses.

The study of the behavior of reinforced soil by a group of columns in axisymmetric condition requires the modeling of reinforcing columns by means of co-centric sand crowns (Figure 6.19) having the equivalent width B_{eq} = 0.18 m

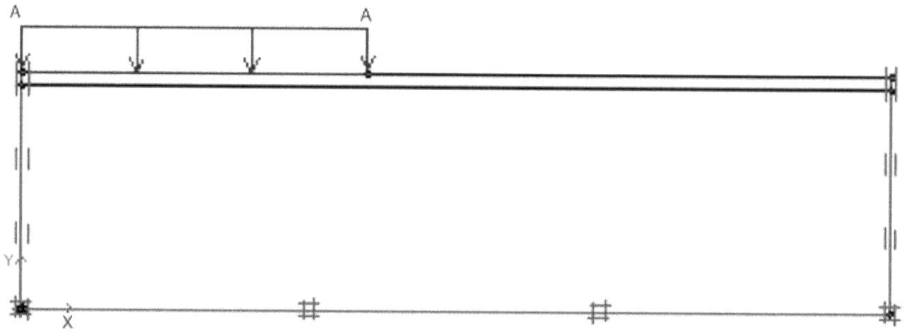

Figure 6.16 Plaxis' model of the La Goulette oil tank in axisymmetric condition

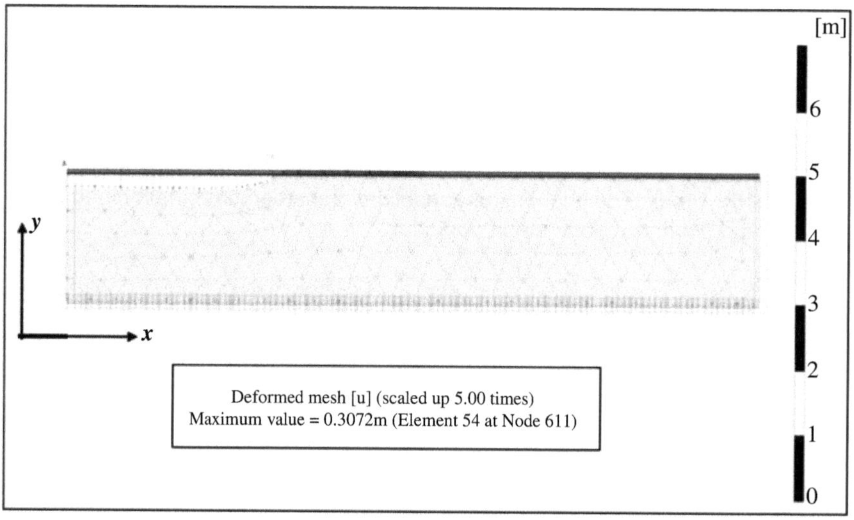

Figure 6.17 Plaxis' mesh of unreinforced soil in axisymmetric condition (La Goulette case study)

and an axis-to-axis spacing equal to 1.5 m. Figure 6.20 shows the modeling by Plaxis software that includes a drainage blanket layer at the surface of reinforced soil. The variation of consolidation settlement at the surface of reinforced soil by sand crowns (axisymmetric study) is plotted in Figure 6.21. From this figure it is noted that the settlement of 9.4 cm at the centerline of the tank is greater than the predicted value of 7 cm in plane strain condition (Figure 6.15). The same remark holds for the predicted settlement at the border of the oil tank: 4.5 and 3.8 cm in axisymmetric and plane strain conditions, respectively.

Table 6.7 summarizes the predicted settlements of consolidation at the centerline of the La Goulette oil tank both in plane strain and axisymmetric studies

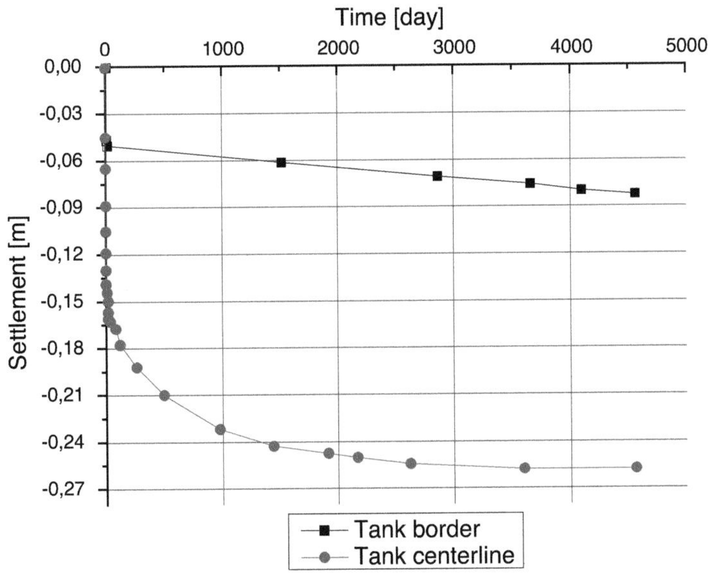

Figure 6.18 Predicted evolution of consolidation settlement at the surface of unreinforced soil (axisymmetric study)

Table 6.7 Predictions of consolidation settlement at the centerline of the La Goulette oil tank using Plaxis V9.2D and FLAC3D codes

Numerical tool	Option computation	Settlement of unreinforced soil (cm)	Settlement of reinforced soil (cm)
Plaxis V9.2D	Plane strain	21.8	7.0
	Axisymmetric study	24.8	9.4
FLAC 3D	3D model	20.25	9.88

performed by the Plaxis V9.2D and 3D modeling generated by the FLAC3D code. Note that the predicted settlements of unreinforced soil are quite similar by the plane strain computations using Plaxis and the FLAC3D codes. In turn, the axisymmetric analysis by the Plaxis code leads to overestimated prediction of consolidation settlement of unreinforced soil. For the case of reinforced soil, it is found that the plane strain analysis slightly underestimates the settlement prediction compared to that obtained by the Plaxis code in axisymmetric condition and the one obtained by the FLAC3D model. We can conclude, based on using any computational numerical tool (Plaxis or FLAC code), that the obtained results on 2D (plane strain and axisymmetric conditions) or truly 3D modeling are almost similar.

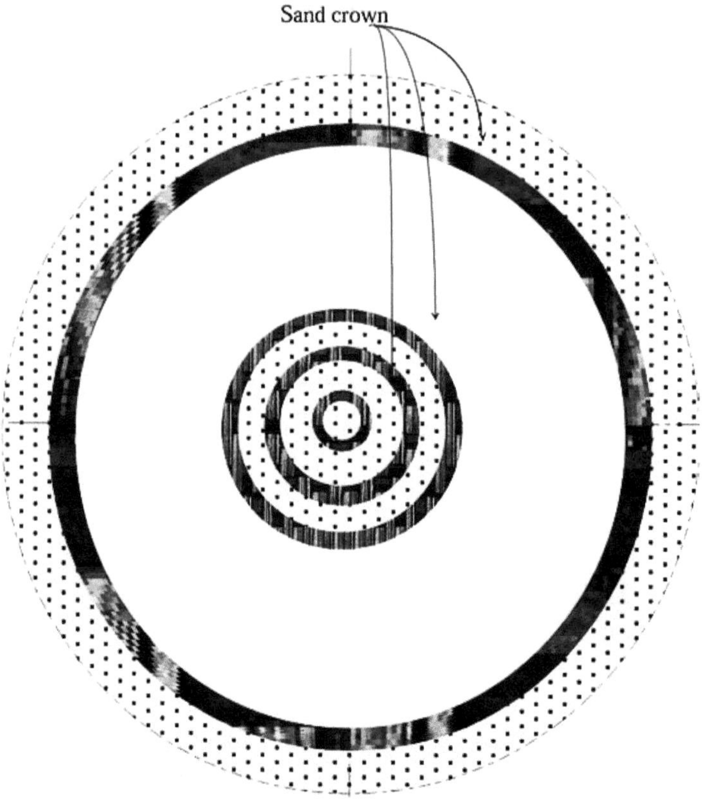

Figure 6.19 Equivalent reinforcement by co-centric sand crowns in axisymmetric condition

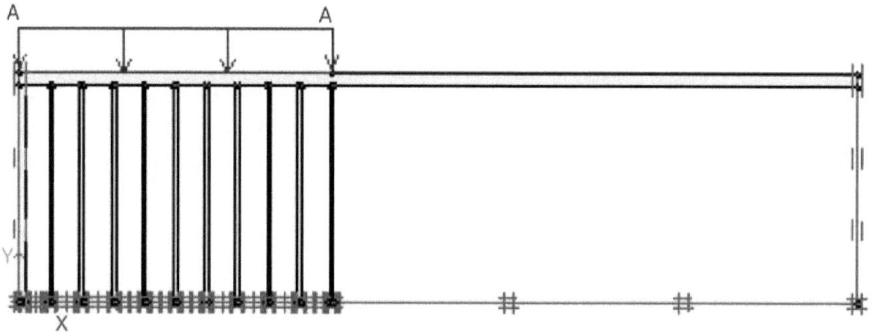

Figure 6.20 Reinforced soil model by Plaxis software in axisymmetric condition

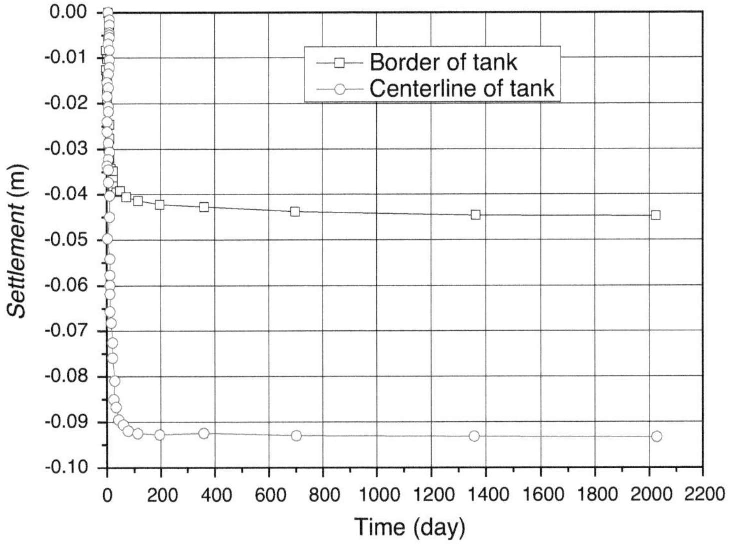

Figure 6.21 Evolution of consolidation settlement at the surface of reinforced soil by sand columns (axisymmetric study)

6.2.3 Embankment on Compressible Soil Reinforced by Floating Stone Columns: Ghannouche Case History (Tunisia)

The Tunisian case history presented in Chapter 3 (§3.6.2) is again reviewed for the study of the behavior of reinforced soil in a post-construction phase. The storage facility consists of two bullets of butane that are 4 m in diameter and 30 m in length, and five bullets of propane that are 5 m in diameter and 45 m in length, protected in mounded banks (Figure 3.10).

The reinforcement by stone columns was considered to be over 11 m in depth. The design of a column-reinforced foundation should comply with the long-term settlement of the bullets' facility that is less or equal to 40 mm over a period of 15 years (El Ghabi et al., 2010).

The behavior of the reinforced ground is studied before and after the stone columns installation, on the basis of analytical and numerical predictions of consolidation settlement. Those predictions will be compared to in situ measured settlement using adequate sensors placed under the embankment.

The soil profile comprises alluvial deposits of silt sand and clay layers (up to approximately 22 m in depth) overlaying a thick, stiff clay layer. The water level is located at a depth of 2.5 m below the ground surface. The geotechnical profile is described in Table 6.8 that shows the average of measured strength from a cone penetration test (CPT) in each layer and the geotechnical properties of compressible layers that are assumed to be normally consolidated.

The construction of the embankment, with a total equivalent vertical load of 120 kPa, was scheduled into two steps as presented in Table 6.9. Referring to the CPT data given in Table 6.8, one can deduce that the applied load is allowable, therefore the stability in terms of bearing capacity is verified. In turn, the stability of the storage facility is rather affected by the non-admissible consolidation settlement, as predicted by Terzaghi's method:

$$\delta_{LT} = \sum_{i=1}^{n} \left(\frac{H_i C_{ci}}{1 + e_{01}} \right) \log \left[\frac{\sigma'_{0,i} + \Delta\sigma_i}{\sigma'_{0,i}} \right] \tag{6.1}$$

Where:

H_i: thickness of compressible layer (i);

$\Delta\sigma_i$: excess of vertical stress induced by the applied load at mid thickness of the compressible layer (i) with normalized compression index "$C_{ci}/1 + e_{0i}$";

$\sigma'_{0,i}$: vertical effective stress calculated at mid thickness of the compressible layer (i).

Using Equation (6.1), the consolidation settlement of unreinforced soil under the embankment is predicted for the two loading stages given in Table 6.9, in three locations shown in Figure 6.22: (1) the centerline, (2) the crest, and (3) the toe.

Table 6.10 indicates that the increase in differential settlement between the centerline and the toe of the embankment passes from 14 cm (at 1st loading stage) to 31 cm (at 2nd loading stage). This differential settlement is very risky for the stability of bullets and the whole embankment as well. Numerical computations in plane strain condition were carried out using the finite element

Table 6.8 Geotechnical profile at Gannouche site including values of CPT strength

Layer	Depth (m)	CPT (MPa)	γ (kN/m³)	e_0	Compression index
Fine sand	0–2.5	12.0	–	–	–
Medium sand with mud and clay inclusions	2.5–9	4.0	–	–	–
Clay and sand partially cemented	9–14.5	30.0	19	0.65	0.1
Alternating bedding of fine sand and clay	14.5–22	30.0	17	1.26	0.395
Mostly hard clay and sand	>22	–	19	0.65	0.1

Table 6.9 Stages of construction of LPG facility

Stage	1	2
Operations	Construction of the first and second layers of the embankment Installation of tanks and hydraulic test	Construction of the third layer of embankment
Time of execution	4 months	5 months

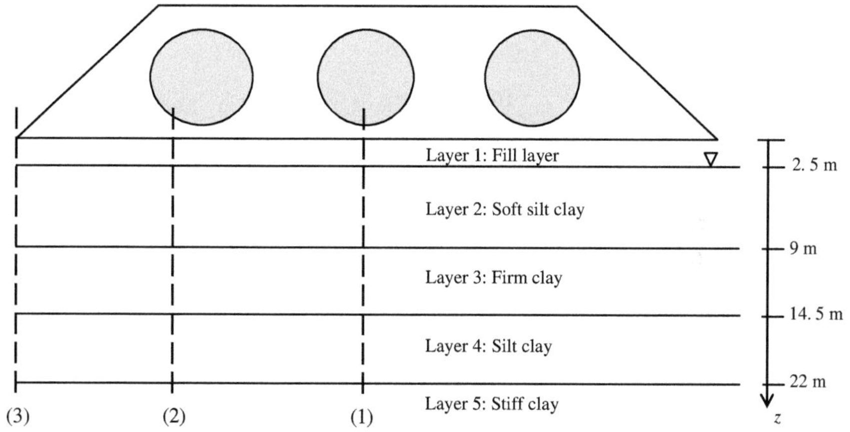

Figure 6.22 Three locations for consolidation settlement predictions under embankment facility (with permission from Editor E. Hamdi)

Plaxis code in order to assess the predicted consolidation settlement by Terzaghi's method. Usage of this tool enabled the analysis of foundation behavior over a fixed period: short, medium, and long terms.

6.2.3.1 Prediction of the Settlement of Unreinforced Soil (Plaxis Code)

Figure 6.23 shows the Plaxis plane strain model of the LPG structure founded on five soil layers of total depth $H_{tot} = 28$ m. The total length and elevation of

Table 6.10 Predicted consolidation settlement of unreinforced soil subjected to full load loading of storage facility (Ghannouche case history)

	Total long-term settlement (cm)	
Location	**First loading stage**	**Second loading stage**
Center line (1)	41	67
Crest (2)	33	61
Toe (3)	27	36

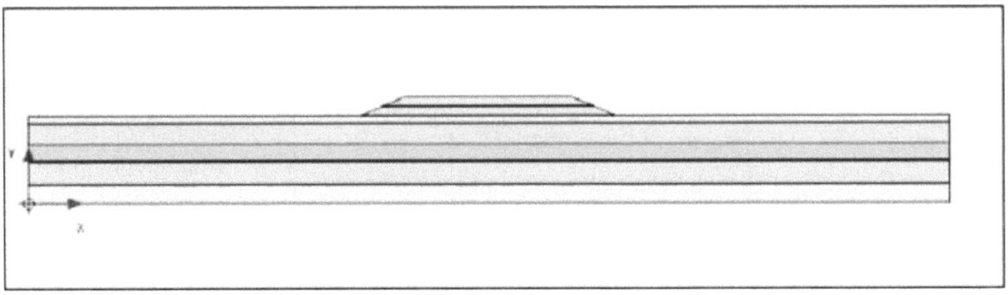

Figure 6.23 Plane strain model of the LPG structure on unreinforced soil

the numerical model are: $L = 129$ m and 34 m, respectively. The LPG structure consists of an embankment of total thickness $H_{emb} = 6$ m; width at base $B = 43$ m. The ratios $\frac{H_{tot}}{H_{emb}}$ and $\frac{B}{L}$ were adopted on the basis of recommendations proposed by Mestat (2001). The constitutive behavior of soil layers and columns material is described by the elastic perfect plastic Mohr-Coulomb model with the parameters given in Table 6.11.

Consider a primary consolidation analysis, with full height of the storage facility, the behavior of the mounded banks is predicted over fifteen years in post construction of the embankment.

The numerical plane strain modeling shown in Figure 6.23 and the one-dimensional Terzaghi's method are different, since those methods use quite different boundary conditions. In fact, unlike the oedometer method that imposes zero horizontal displacement, the Plaxis code that incorporates both the horizontal and vertical consolidation takes into account the horizontal displacements of the ground and, then, it leads to a more realistic prediction of consolidation settlement. Figure 6.24 shows the contours of vertical displacements within the embankment and unreinforced soil layers.

Table 6.11 Geotechnical parameters of soil layers, columns, and embankment material adopted in plane strain numerical analysis (Ghannouche case history)

	Height/ Thickness [m]	Young's Modulus [MPa]	Cohesion [kPa]	Friction angle [°]	Total unit weight [kN/m^3]
Backfill material	6	10	1	30	20
Fine sand	2.5	30	5	30	19
Soft silt clay	6.5	5.7	2	24	18
Firm clay	5.5	60	2	24	19
Silt clay 1	7.5	12	15	10	18
Stiff clay 2	6	80	2	24	20

The variation of settlement of unreinforced soil versus time at the centerline and crest of the embankment is plotted in Figure 6.25. It indicates non-admissible differential settlement between the centerline and crest locations.

As depicted in Figure 6.25, the predicted consolidation settlement of soft silt clay layers is too significant. The maximum predicted differential settlement (difference between predicted values at the axis and toe of the embankment) is 8 cm—that greatly exceeds the allowable limit of 4 cm.

Since the soil foundation would undergo non-admissible differential settlement threatening the stability and durability of the storage facility, floating stone columns were installed beneath the embankment, up to 11 m in depth. This depth is justified from Figure 6.24 which demonstrates that the settlement becomes less than one centimeter beyond 11 m depth. Therefore, there is no need to install columns exceeding 11 m in length to reduce the settlement.

Adding to the benefits in increasing the bearing capacity and settlement reduction, the use of stone columns contributes in the acceleration of primary consolidation since columns are made up of gravel material having high permeability; therefore, they behave like vertical drains by accelerating the horizontal drainage within the compressible layers.

6.2.3.2 Numerical Simulation of the Behavior of a Storage Facility on Reinforced Soil

The numerical model proposed in Chapter 5 (§5.3.2) is again considered to predict the behavior of reinforced soil that resulted from the installation of

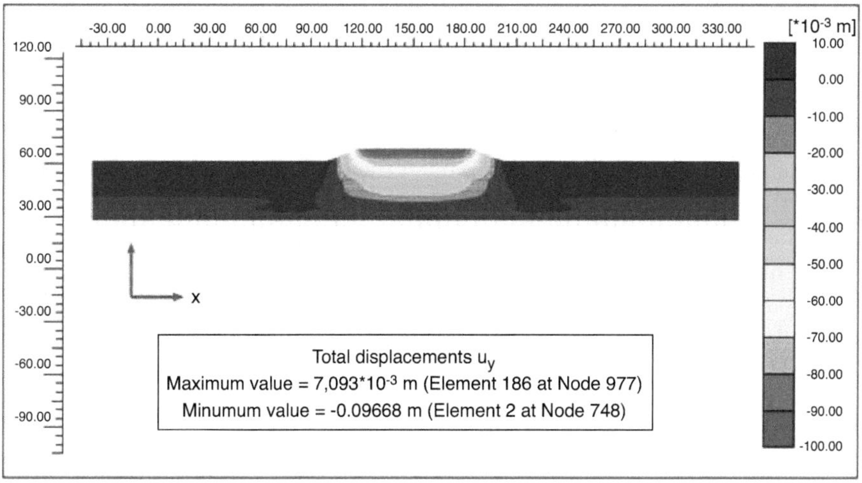

Figure 6.24 Contours of vertical displacements of embankment facility on unreinforced soil subjected to uniform load of 120 kPa

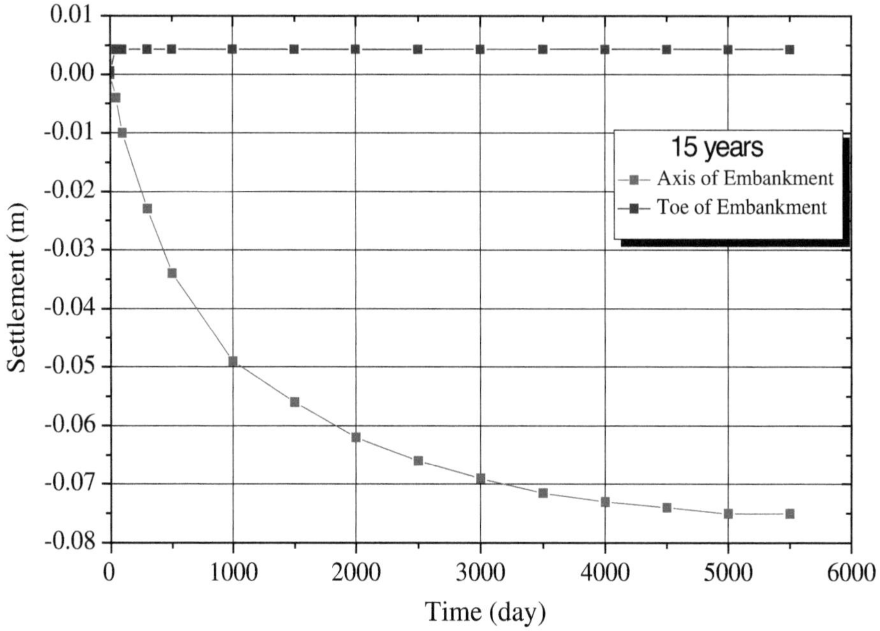

Figure 6.25 Variation of consolidation settlement vs. time of unreinforced soil by Plaxis code

seven stone columns in a triangular pattern, as depicted in Figure 5.15. Hence, a numerical study was undertaken using an axisymmetric model of a stone column surrounded by a crown of stone (that is equivalent to six stone columns) shown in Figure 6.26 (Ellouze and Bouassida, 2009). The equivalent thickness of the stone's crown b_{cr} is calculated from Equation (5.3).

The soil is reinforced by end-bearing vibro-displacement stone columns of 0.86 m in diameter, installed in 2.2 m triangular grid spacing. The geotechnical parameters of soil layers, embankment material (Table 6.11), and columns material are all governed by the Mohr-Coulomb constitutive law. Table 6.12 displays the parameters of columns material.

The numerical procedure *dummy material* is performed to simulate the lateral expansion of stone material both for the central column and the equivalent stone's crown, as detailed in Chapter 5 (§5.3.2). By performing this numerical procedure, the horizontal displacement is increased up to the final column's radius $a = 0.43$ m and the equivalent crown's thickness of $b_{cr} = 0.25$ m are attained (Bouassida et al. 2013).

The installation of the two fill layers and tank uniformly distributed load of 60 kPa is simulated and followed by a consolidation procedure during four

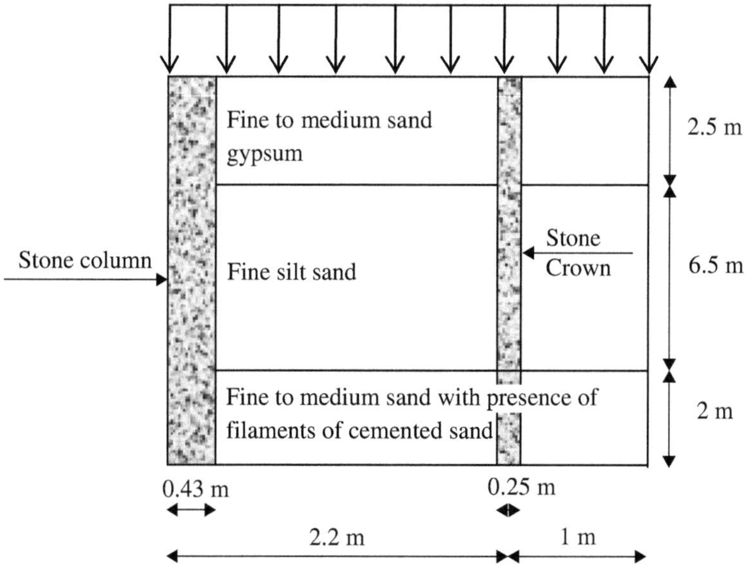

Figure 6.26 Axisymmetric model of soil reinforced by a group of stone columns (Ghannouche case study)

Table 6.12 Geotechnical parameters of columns material of the numerical model in plane strain condition (Ghannouche case history)

	Length [m]	Young's Modulus [MPa]	Cohesion [kPa]	Friction angle [°]	Total unit weight [kN/m^3]
Stone material	11	60	0.005	40	20

months. Then, the uniformly distributed load was increased up to 120 kPa at the surface of the reinforced soil model (Table 6.9), as scheduled for the step of installation of the last fill layer. The final stage of numerical computation, using the Plaxis V9.2D code, comprises a primary consolidation during 15 years.

Results

Using the UCM shown in Figure 6.26, the predicted settlements of unreinforced soil for the first and second stages of construction and the consolidation settlement after 15 years of consolidation were 6 cm and 12 cm, respectively. Figure 6.27 confirms that those settlements are constant at the surface of the adopted UCM when considered for the unreinforced soil. The magnitude of settlement at the embankment axis, as predicted by the plane strain model shown in Figure

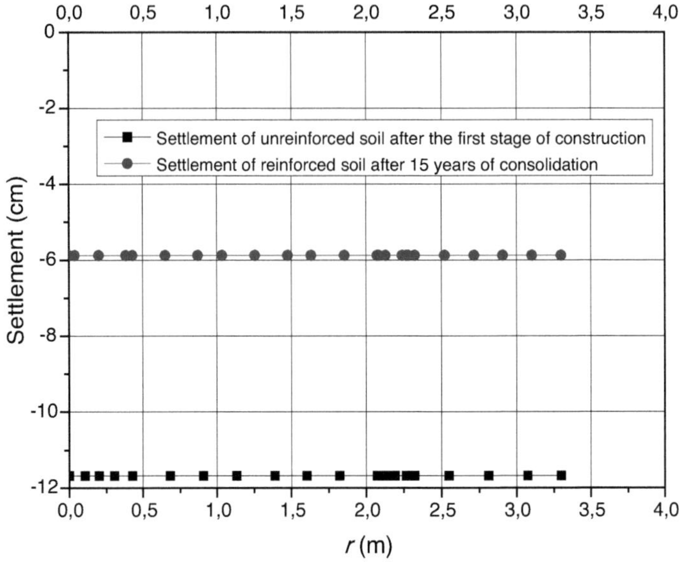

Figure 6.27 Prediction of consolidation settlement of unreinforced soil by the UCM

6.24, is lower than predicted by the UCM. This gap in predictions is attributed to the difference between the plane strain and axisymmetric models, as well as to the difference between the applied load for each model.

The predicted settlement of reinforced soil is close to 3.5 cm during the first stage of construction (4 months of consolidation); then it reaches 4.5 cm for the long-term analysis of 15 years after the completion of embankment construction (Figure 6.28). The factor of settlement reduction increased from 1.67 to 3 from the first stage of embankment construction to the final loading stage.

6.2.3.3 Numerical Predictions Using the Group of Trenches Model

The plane strain model comprises a 6 m height embankment found on the soil profile shown in Figure 6.29.

Forty six stone columns were installed along the horizontal direction, with axis-to-axis spacing of 1.9 m, and along the perpendicular direction, 30 stone columns were installed with axis-to-axis spacing of 2.2 m over 64 m in length.

The group of stone columns is modeled by a group of equivalent trenches to simulate the behavior of reinforced ground in plane strain condition using the Plaxis code.

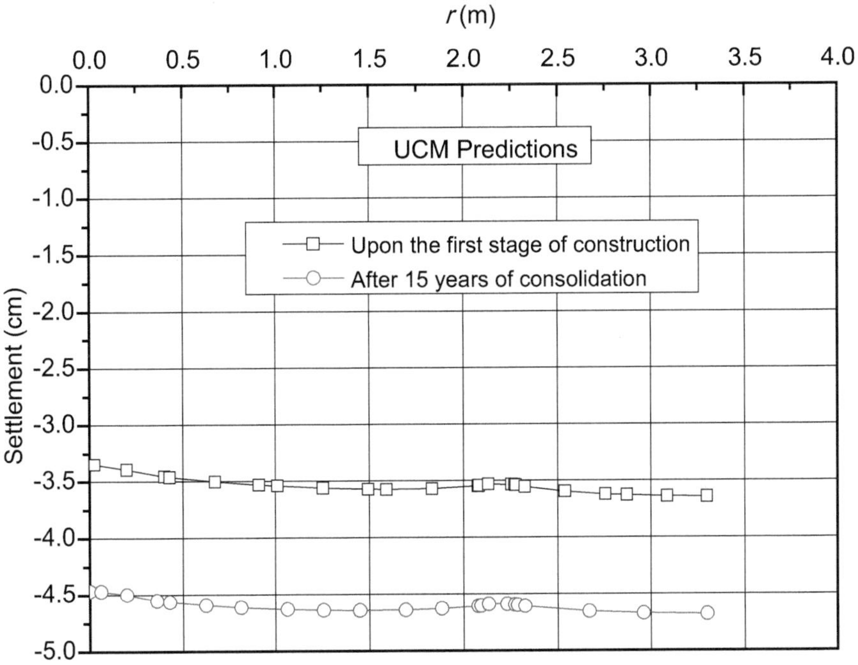

Figure 6.28 Prediction of consolidation settlement of reinforced soil by the UCM

Then, the equivalent thickness of a trench, b_{tr}, is determined from Equation (6.2):

$$b_{tr} = \frac{30}{64}\pi a^2 \tag{6.2}$$

Forty one trenches made up of stone material are considered in the numerical model with dimensions: thickness, b_{tr} = 0.3 m; length, H_c = 11 m; spacing between edges of trenches, s' = 1.9 m (Klai et al., 2015).

The behavior of soil layers and columns material is described by the elastic perfect plastic Mohr-Coulomb constitutive law with parameters given in Tables 6.11 and 6.12, respectively.

Using the plane strain modeling shown in Figure 6.29, the numerical simulation of an embankment with staged construction comprises four phases. The first phase consists of 30 days of consolidation analysis that occurs upon the installation of stone columns within the compressible layers. Then, a partial horizontal consolidation is triggered that dissipates the induced excess pore

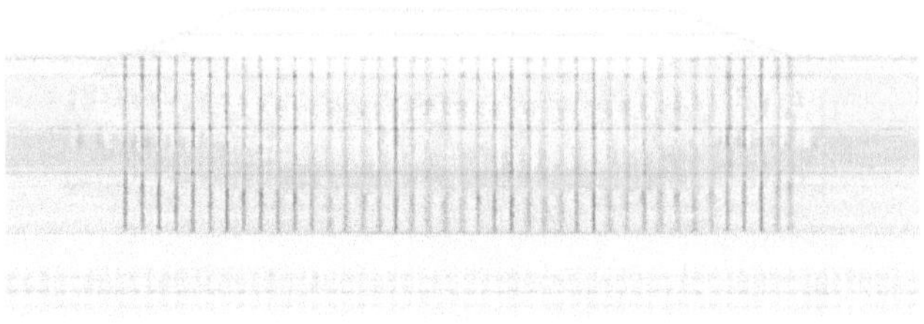

Figure 6.29 Geometry of numerical plane strain modeling of reinforced soil (Ghannouche case study)

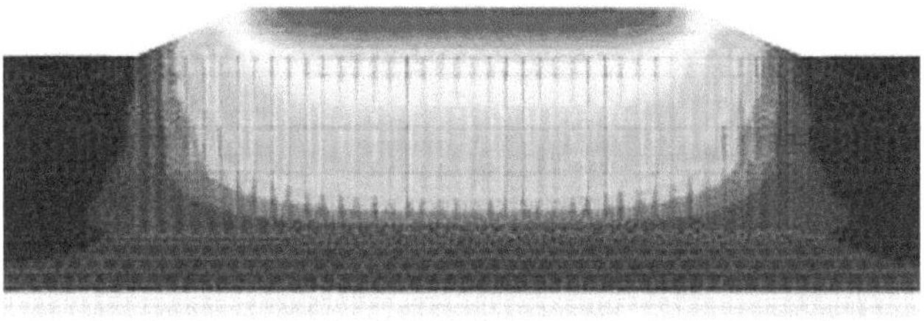

Figure 6.30 Contours of vertical displacements of the embankment facility on reinforced soil by stone columns subjected to a uniform load of 120 kPa

pressure and leads to uniform consolidation settlement within the compressible layers. The second and third phases of numerical staged construction also consist of consolidation analysis which corresponds to loadings applied upon the execution of first and second embankment layers, respectively. A consolidation analysis is run for each loading embankment layer during 120 days and 270 days, respectively.

Figure 6.30 shows that the contours of vertical displacement with a maximum value of settlement equals 8 cm at the upper crest of the embankment facility. While at the surface of reinforced soil, the predicted settlement is an almost uniform value of 6 cm over the width of the upper crest beneath the embankment. Hence, under the bullets of the mounded bank schematized in Figure 6.22, there is no differential settlement.

Figure 6.31 shows the variation of consolidation settlement versus time. This settlement is stabilized at a magnitude of 6.5 cm for up to four years.

The follow-up of behavior of a storage facility built on reinforced ground by floating stone columns was performed by means of a data acquisition unit connecting the pressure sensors, to record the evolution of settlements, located at the surface of reinforced ground as shown in Figure 6.32.

The measured settlements resulted from the variation of elevation between reference sensors PR08CA1 and PR08CA2 and the sensors placed in each

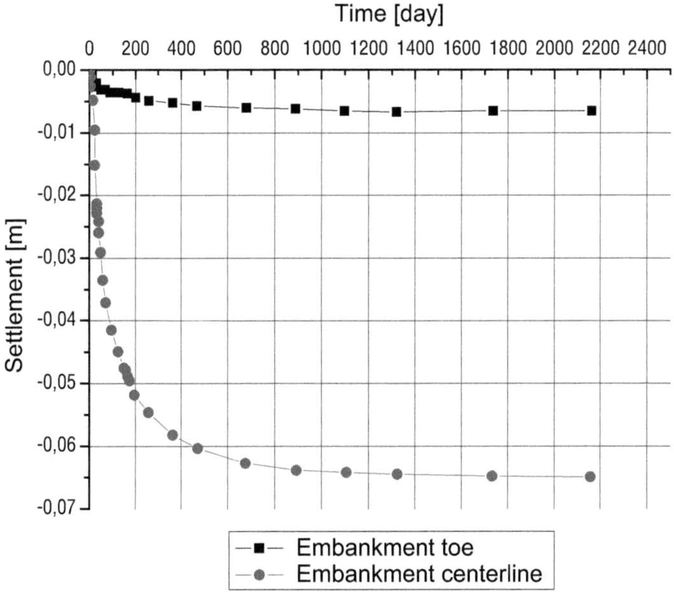

Figure 6.31 Variation of consolidation settlement vs. time of reinforced soil by floating stone columns at the centerline and border of the embankment facility

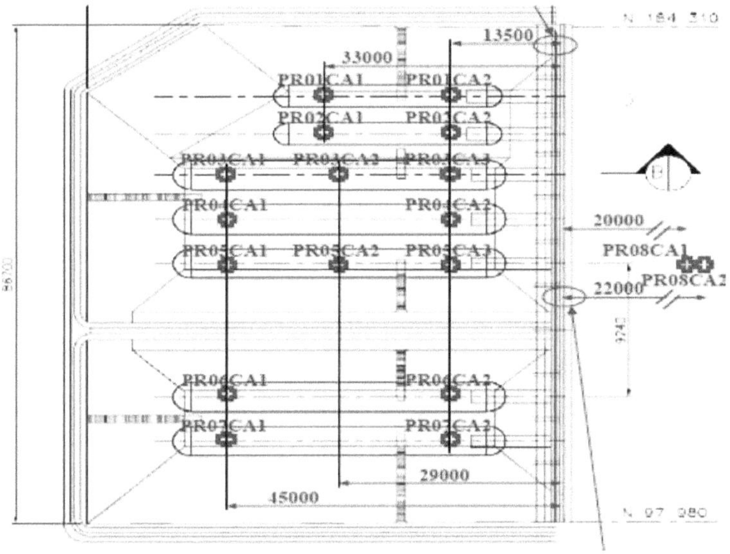

Figure 6.32 Location of sensors (dimensions are expressed in mm) for the follow-up of settlement under the embankment facility (with permission from Editor E. Hamdi)

profile (two to three sensors under each bullet). All sensors were connected by a hydraulic network where the variation of water level corresponds, thus, to the measured variation of the hydraulic load of the liquid. More details about the topographical survey are given in El Ghabi et al. (2010).

The first measured settlements induced by the acquisition unit data occurred in post-construction of the backfill, and then stabilized when the data acquisition measurements started. The evolution of measured settlements in function of time at profiles PR02, PR03, PR04, and PR05 are plotted in Figures 6.33, 6.34, 6.35, and 6.36, respectively. Those measured settlements after the installation of the first backfill layer varied between 1.0 and 1.7 cm, and then were stabilized after a period of 30 days. After the completion of the entire embankment, the magnitude of measured settlements was less than 3.0 cm (El Ghabi et al., 2010). Based on this observation it is confirmed that the stone columns reinforcement experienced at the Gannouche site fulfilled the requirement of admissible consolidation settlement that was less than 3.5 cm.

Figures 6.33, 6.34, 6.35, and 6.36 also indicate that the measured settlement reached about 6 cm under the crest of the embankment in post-construction of the final embankment layer. Indeed, a similar trend of the consolidation settlement has been predicted by Plaxis code for the same elapsed time of construction (Figure 6.30). Indeed, the maximum computed consolidation settlement is at 6.5 cm under the crest of the embankment.

The last phase of numerical staged construction consisted in consolidation analysis that corresponds to the final height of the embankment. This phase simulates the long-term behavior 15 years after the construction of the storage facility. Note that during the progress of stage construction of the embankment, the settlement significantly increases in the different locations, and then it becomes almost uniform within the allowable limit of settlement that is 4 cm. This long-term settlement corresponds to the induced deformation within unreinforced sublayers. After Bouassida and Hazzar (2015), the settlement of the reinforced layer depending on the length of columns, accelerated by the drained columns' material, takes place during embankment construction.

6.2.4 Comparison Between Predictions by the Plane Strain and Axisymmetric Model of Reinforced Soil

Using the UCM shown in Figure 6.26, the settlement prediction after four months of consolidation is in accordance with measured settlements at profiles PR02, PR03, PR04, and PR05.

While the consolidation settlement (after 15 years of consolidation) as predicted by the UCM is lower than that measured under the embankment, the follow-up of settlement evolution was limited to 14 months. This gap of

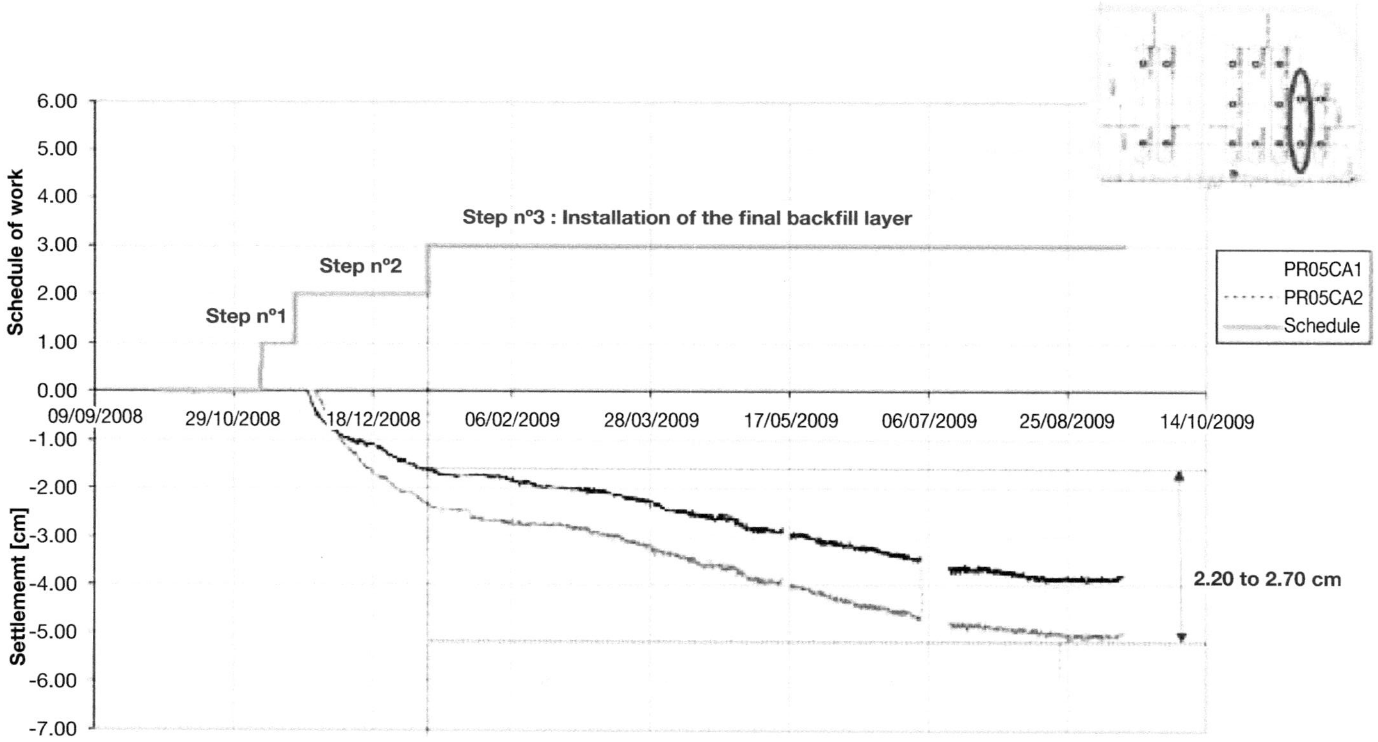

Figure 6.33 Evolution of measured settlement vs. time at location PR02 (with permission from Editor E. Hamdi)

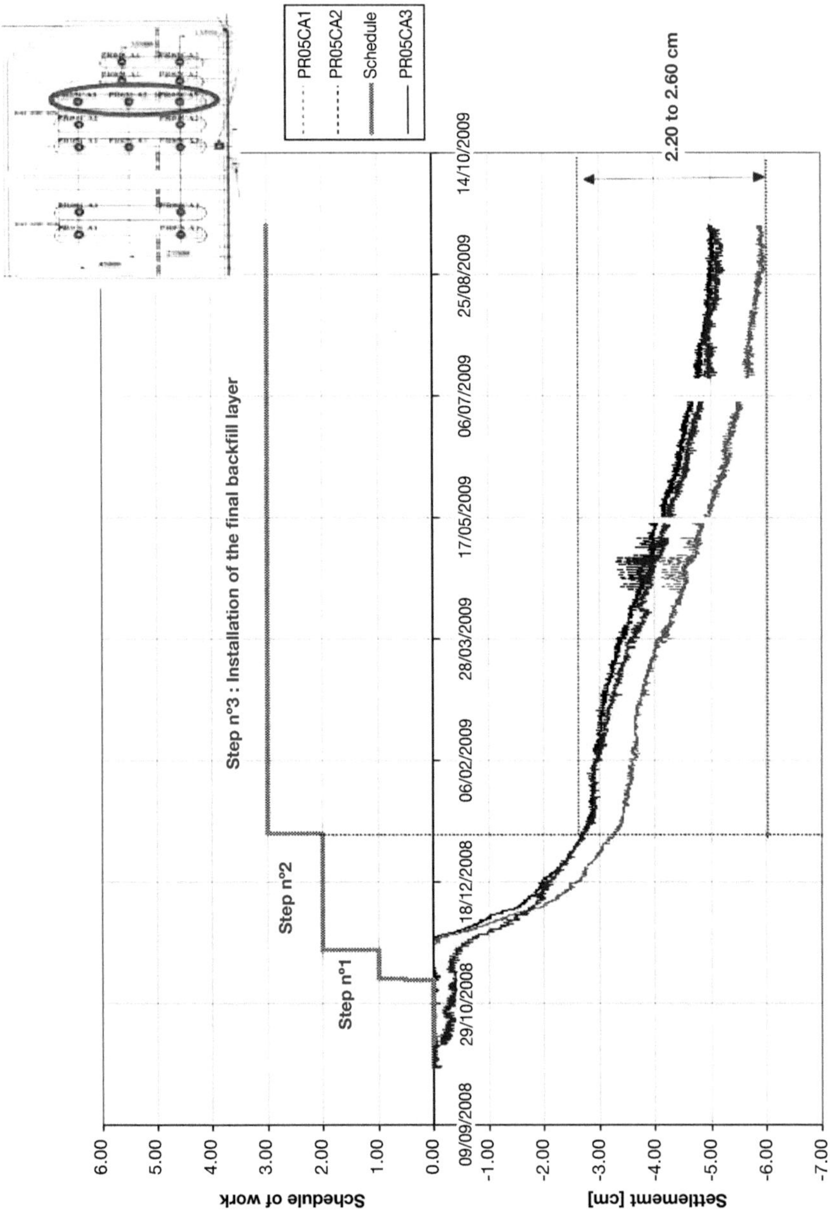

Figure 6.34 Evolution of measured settlement vs. time at location PR03 (with permission from Editor E. Hamdi)

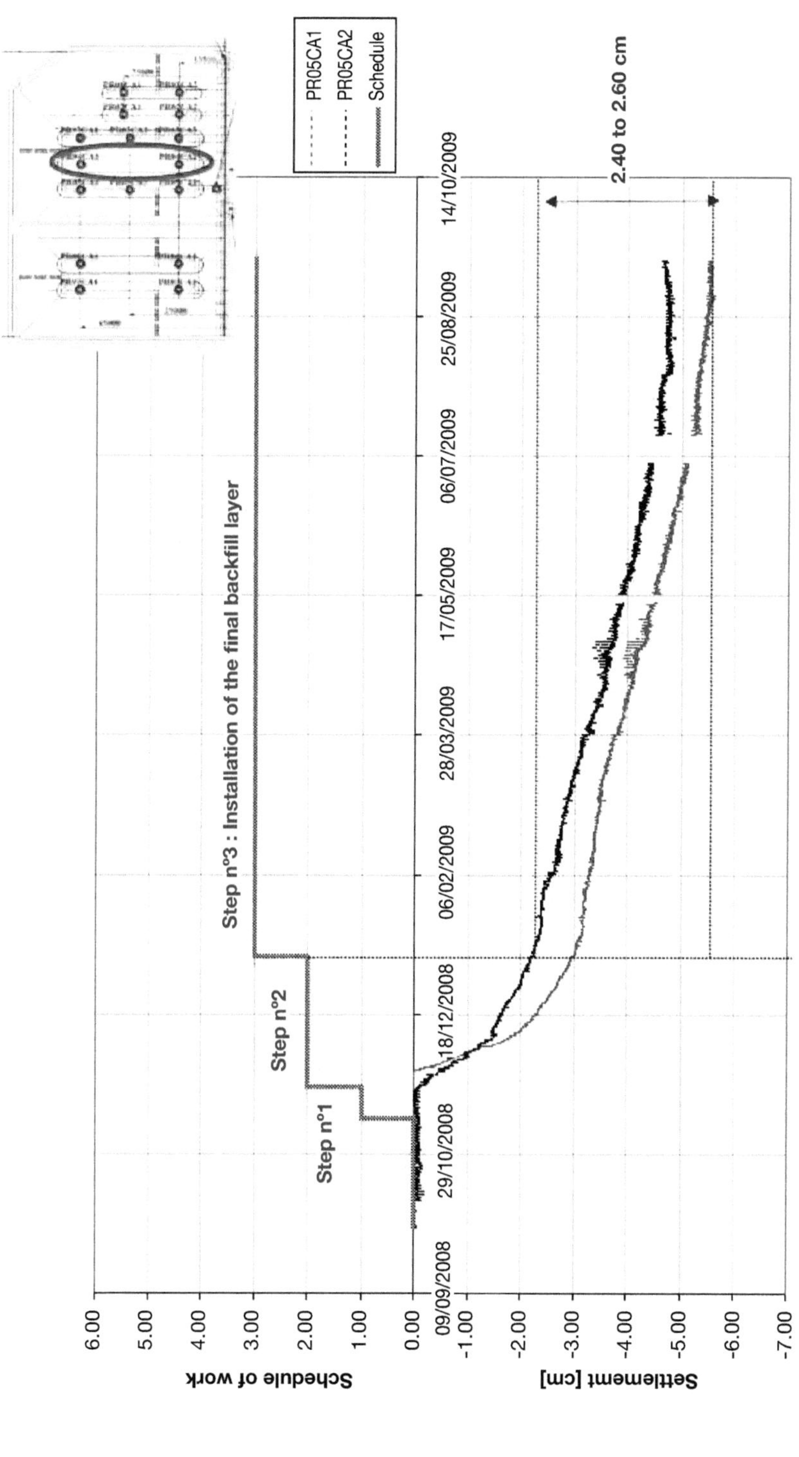

Figure 6.35 Evolution of measured settlement vs. time at location PR04 (with permission from Editor E. Hamdi)

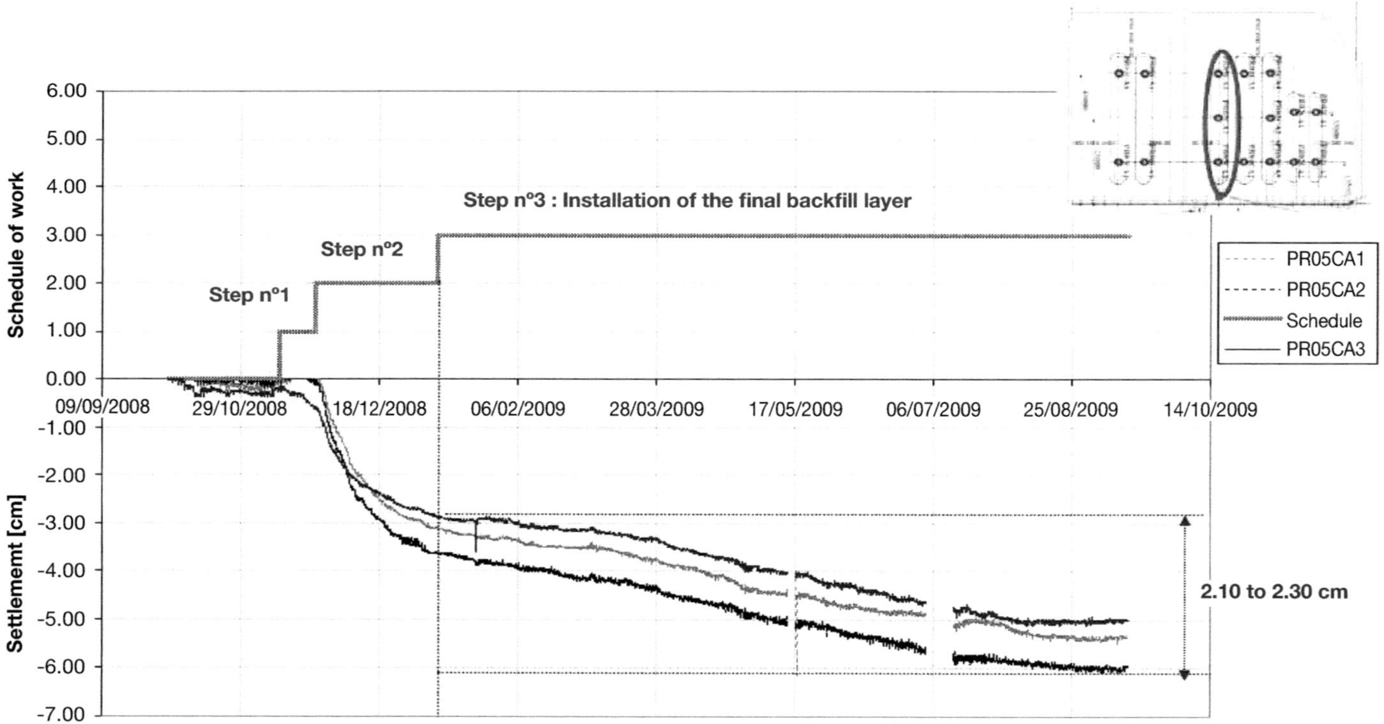

Figure 6.36 Evolution of measured settlement vs. time at location PR05 (with permission from Editor E. Hamdi)

consolidation settlement is owed to the constrained horizontal displacement at the border of the UCM. Nonetheless, it is also noted that the presence of the blanket stone layer at the surface of reinforced soil was not considered by the UCM.

The study of the Ghannouche case history demonstrated the usefulness of floating stone columns reinforcement, as no residual settlement occurred in the unreinforced sublayers. It is then concluded that the design of a foundation of bullets of butane and propane (integrated into an embankment) on compressible layers reinforced by stone columns was successful. Indeed, this design complied with the allowable settlement of the foundation over 15 years, as predicted by the numerical computations, in plane strain condition using the finite element Plaxis code. Those predictions revealed an acceptable agreement with the measured settlement that remained under 4 cm over 15 years.

6.3 CONCLUSIONS

This chapter focused on the study of the behavior of foundations on compressible soils reinforced by sand and stone columns. Three Tunisian case studies have been investigated by performing numerical computations using the Plaxis V9.2D and the FLAC3D codes. Main findings are summarized as follows:

1st Case Study

All methods of prediction consider linear elastic behavior both for the initial soil and column material. Note that predicted settlement by the FLAC3D code is close to that obtained by the analytical method proposed by Bouassida et al. (2003), which adopts the group of columns model that is similar to the FLAC3D code. The relative difference between those predictions, 0.59% vs. 3.74%, for the unreinforced soil and reinforced soil respectively, are negligible. This first example confirms the usefulness of analytical methods of settlement prediction programmed within the *Columns 1.01* software which is based on the optimized IAR.

2nd Case Study

It is important to note that the difference between predicted settlements of unreinforced soil is not significant; therefore, the use of any computational tool (Plaxis or FLAC) or any computation option (either plane strain or axisymmetric study) leads to comparable results from which the reinforcement by columns can be justified. In turn, when dealing with the modeling of reinforced soil, it revealed that the plane strain condition slightly underestimates the settlement prediction. Indeed, such hypothesis is not realistic for cylindrical structures for which the axisymmetric condition can be fairly adopted since its settlement prediction is close to that obtained by 3D modeling.

3rd Case Study

The design of a foundation of bullets of butane and propane on compressible layers reinforced by floating stone columns was successful. Indeed, this design complied with the allowable settlement of the foundation over 15 years, as predicted by the numerical computations, in plane strain condition, using the Plaxis code. Those predictions were revealed to be in good agreement with the measured settlement of less than 4 cm over 15 years.

FORTHCOMING DEVELOPMENTS

Throughout this book, the design of foundations on soils reinforced by columns has been deeply investigated in regard to the most viewed benefits, e.g., bearing capacity, settlement, and acceleration of consolidation. More recently, other investigations have been published on this subject—for example, the performance of stone column–reinforced soft ground under static and cyclic loads (Basack et al. 2015). In parallel, at the Research Laboratory of Geotechnical Engineering of the National Engineering School of Tunis, the study of liquefaction mitigation by stone columns also started. Ben Salem et al. (2015) reported the primary results regarding the effect of granular column installation on excess pore pressure variation during soil liquefaction. Those very recent published works represent a new trend related to the design of column-reinforced foundations.

REFERENCES

Basack, S., Indraratna, B. and Rujikiatkamjorn, C., Modeling the Performance of Stone Column–Reinforced Soft Ground Under Static and Cyclic Loads. *J. Geotech. Geoenviron. Eng.*, 142 (2), 2016. 04015067: 1-15.

Ben Salem, Z., Frikha, W. and Bouassida, M., "Effect of Granular-Column Installation on Excess Pore Pressure Variation During Soil Liquefaction." *Int. J. Geomech.*, 16 (2), 2016, 04015046. 1-8.

Bouassida, M., Guetif, Z., de Buhan P. and Dormieux, L., Estimation par une approche variationnelle du tassement d'un sol renforcé par colonnes. *Revue Française de Géotechnique*. 102 (1), 2003, 21–29.

Bouassida, M. and Hazzar, L., Novel tool for optimized design of reinforced soils by columns. Ground Improvement: Proc. ICE, London 165 (1), 2012, 31–40.

Bouassida, M., Ellouze, S., Ben Salem, Z. and Znaidi, M. N., Numerical study of the behavior of improved soft clay by stone column. Keynote lecture, Proc. Foundation and Soft Ground Engineering Conference. Thu Dau Mot University ICTDMU-1, June 5–6, Binh Duong, Vietnam, Hai et al. Eds, 2013, Construction Publishing House, 1, 55–66.

Bouassida, M. and Carter, J. P., Optimization of Design of Column-reinforced Foundations. *Int. J. Geomech.*, 14 (6), 2014, 04014031-1-10.

Bouassida, M, Klai, M, Tabchouche, S. and Mellas, M., On the behavior of columnar-reinforced foundations. Proc. 16th A.R.C.S.M.G.E, Hammamet (Tunisia), 26–29 April 2015, 359–364.

Bouassida, M. and Hazzar, L., Performance of soft clays reinforced by floating columns. Book "Ground Improvement Cases Histories, Embankments with Special Reference to Consolidation and Other Physical Methods. Editors Indraratna et al., Chap 16. Part Two: Sands and Gravel Piles, Stone Columns and Other Rigid Inclusions. Butterworth Heinemann publications. 2015 Elsevier, 433–449.

Chai, J. C., Shrestha, S., Hino, T., Ding, W. Q., Kamo, Y. and Carter, J. P., 2D and 3D analyses of an embankment on clay improved by soil–cement columns. *Computers and Geotechnics* 68, 2015, 28–37.

Ellouze, S. and Bouassida, M., Prediction of the settlement of reinforced soft soil clay by a group of stone columns. Proc. 2nd Int. Conf. New Develop. in SMGE. Edit. Atalar et al. 28–30 May 2009, N.E.U. Nicosia (Cyprus). 182–187.

El Ghabi, B., Hermges, T., Lambert, S., Ellouze, S., Bouassida, M. and Mejri, A., Comparison between predicted settlement of a group of stone columns foundation and in situ records. 2nd Int. Conf. Geotech. Eng. October 25–27, 2010, Tunisia, 255–264.

Han, J. and Ye, S. L., A simplified method for computing consolidation rate of stone column reinforced foundations. *Journal of Geotechnical and Geoenvironmental Engineering*, ASCE, 127(7), 2001, 597–603.

Indraratna, B., Basack, S. and Rujikiatkamjorn, C., Numerical solution to stone column reinforced soft ground considering arching, clogging and smear effects. *J. Geotech. Geoenv. Engrg.*, 139 (3), 2013, 377–394.

Klai, M., On the behavior of Tunis soft clay—Application to the study of foundations stability (in French). Defended 16 Oct. 2014, National Engineering School of Tunis, Tunisia.

Klai, M., Bouassida, M. and Tabchouche, S., Numerical modelling of Tunis soft clay. *Geotechnical Engineering Journal of the SEAGS & AGSSEA*, 46 (4), 2015, 87-95.

Mestat, P., Momis: une base de données sur la modélisation numérique des remblais sur sols compressibles et sur la confrontation calculs-mesures in situ. Bulletin de Liaison du Laboratoire Central des Ponts et Chaussées (France), 2001, 202, 43–58.

Sexton, B. G. and McCabe, B. A., Numerical modeling of the improvements to primary and creep settlements offered by granular columns. *Acta Geotechnica*, 2013, 8:447–464.

Six, V., Mroueh, H., Shahrour, I. and Bouassida, M., Numerical Analysis of Elastoplastic Behavior of Stone Column Foundation. *Geotech Geol. Eng. J.* 2012, 30:813–825.